Tereza Cristina Buonocore Nunes

Training of practical instructors of air traffic controllers

Tereza Cristina Buonocore Nunes

Training of practical instructors of air traffic controllers

Operational safety

ScienciaScripts

Imprint

Any brand names and product names mentioned in this book are subject to trademark, brand or patent protection and are trademarks or registered trademarks of their respective holders. The use of brand names, product names, common names, trade names, product descriptions etc. even without a particular marking in this work is in no way to be construed to mean that such names may be regarded as unrestricted in respect of trademark and brand protection legislation and could thus be used by anyone.

Cover image: www.ingimage.com

This book is a translation from the original published under ISBN 978-3-639-75892-4.

Publisher:
Sciencia Scripts
is a trademark of
Dodo Books Indian Ocean Ltd. and OmniScriptum S.R.L publishing group

120 High Road, East Finchley, London, N2 9ED, United Kingdom
Str. Armeneasca 28/1, office 1, Chisinau MD-2012, Republic of Moldova, Europe
Printed at: see last page
ISBN: 978-620-5-81230-3

DEDICATION

I dedicate this work to my mother, who always made me believe in myself and feel special enough to achieve my dreams. While she was by my side, she gave me strength to overcome challenges and in her absence, strength to never give up. To my father for the example of the life I decided to follow. To my husband for his understanding, support, care and love during the development of this work. And to my dear daughter, who gave me a new meaning to life.

THANKS

I thank God first and foremost for always being by my side.

I thank Col. Av. Ricardo Barion for the opportunity to take the course, for his trust and friendship.

To Prof.ª Lígia, for going beyond her role as advisor and believing in me, despite the adversities, and giving me emotional support to pursue the development of the work in the difficult moments I went through.

To Professor Donizeti for the opportunity and trust.

To the friend and co-supervisor Cap. Jorge Augusto who gave me his precious knowledge in the area of air traffic control and great experience of instruction in the Air Force

SUMMARY

Improving the training of flight controllers is very important to strengthen air transport in Brazil. With the increasing flow of aircraft it is of utmost importance to have a well-trained professional to perform the function in order to ensure safety in the operation.

The training of this professional goes through distinct stages, one theoretical and the other practical. Although closely linked and interdependent, they occur at different times. The practical phase is subdivided into two stages, simulated practice and real practice. The real one occurs with the professional, still in training, already at his workstation with the monitoring of a practical instructor, responsible for monitoring and training the trainee who was approved in the previous phases and is already able to train in real time.

The practical instructor is of fundamental importance in air traffic controller training and assumes a huge responsibility when dealing with professionals still in training during operation. This instructor aims to bring new knowledge to the trainees, make them work with unusual situations, act with naturalness, objectivity and assertiveness in the exercise of the function, identify possible doubts or learning problems and correct them along the teaching process, monitoring actions and constantly evaluating the trainee to ensure the quality of the air traffic control professional's training, minimizing risks to aviation regarding control. The function of this instructor is very relevant and requires, in turn, an adequate training with specific focus for performance in practical instruction.

In this context, the objective of this work is to propose an adequate training for these professionals in line with international guidelines. For this purpose, a research of qualitative nature was conducted, using mainly academic and documentary literature review to propose a basic curriculum for the training of the practical air traffic controller instructor. The research was structured in order to present and analyze the national and international panoramas regarding the characteristics of the courses offered and content of the practical-operational air traffic controller (ATC) instructor training. A primary data search was also conducted to identify air traffic problems that could be attributed to problems in instruction. Additionally, it was presented and discussed the PHO that gathers the main improvements that have been developed to improve operator instruction. This program is an advance, but we identified the need for the instructor to have previous knowledge so that the instructor's operational qualification is successful. Finally, a chapter is dedicated to propose a set of basic pedagogical contents to be taken into consideration when building the practical instructor qualification, thus forming part of a curricular base able to strengthen the controllers' instructors training.

ABSTRACT

The improvement on the training of air traffic controllers is very important in order to strengthen air transport in Brazil. With a gradual increase in the flow of aircraft, it is of vital importance to have a professional who is fully prepared to perform his duties so as to ensure safety in the operations.

The training of such professional goes through distinctive stages, a theoretical and a practical one. Although both are intrinsically connected and interdependent, they occur in different moments. The practice is divided into another two stages, the simulation practice and the real practice. The real practice occurs when the professional, who is still under training, is already in his work station accompanied by a practice instructor responsible for this trainee who has been approved in the previous stages and is found suitable for real time training.

The practice instructor is of the essence in the formation of an air traffic controller and takes on a lot of responsibility when dealing with professionals who are still under training during the operations. Such instructor aims at passing his knowledge on to the trainees, allowing them to work in non-routine situations with naturalness, objectiveness and assertiveness while performing their duties, being able to identify possible doubts or problems derived from their training and correct them throughout this process by means of monitory and by constantly evaluating the trainee so as to ensure the quality in the training of air traffic controllers, this way minimizing the risks to aviation regarding air traffic control. The function of this instructor is extremely relevant and requires appropriate training with specific focus on practice instruction.

In this context, the aim of this work is to suggest an appropriate training for these professionals in consonance with international guidelines. For this, a qualitative research has been carried out by means of reviewing academic and documental literature so as to suggest a basic curriculum for the formation of a practice instructor to air traffic controllers. The research has been structured so as to present and analyze the national and international view on the characteristics of the courses offered as well as the content of trainings aimed at practical-operational instructors to air traffic controllers. A search in primary data has been conducted in order to identify the problems in air traffic that could be attributed to problems in instruction. Moreover, the Operational Certification Programme (PHO) has been presented and discussed, which gathers the main improvements that have been developed in order to improve the instruction of the operator. Such programme is a step forward; however, it has been identified the instructor's need for previous knowledge in order to ensure that the instructor's operational certification is successful. Finally, a whole chapter has been dedicated to suggest a set of basic pedagogical content to be taken into consideration in the formation of practice instructors, leading to a curriculum basis able to improve the training of air traffic controllers' instructors.

SUMMARY

LIST OF ABBREVIATIONS

ACC - Area Control Centre;

ACOP - Operational Agreement;

APP - Approach Control;

AOO

ASEGCEA - Brazilian Airspace Control System Operational Security Advisory;

ATCO - Air Traffic Controller;

ATS - Air Traffic Service;

AVOP - Operational Notice;

CINDACTA - Integrated Centre for Air Defence and Air Traffic Control;

CM - Minimum curriculum;

CPI - Air Force Instructor Preparation Course;

CTP001 - SISCEAB Instructors Standardization Course until (2007) and SISCEAB Instruction

Standardization Course from 2007 onwards;

CTP006 - Training Course for SISCEAB Practical Operational Trainers;

CT - Curitiba;

CTA - Air Traffic Control;

CTA - Air Traffic Controller;

DECEA - Department of Airspace Control;

DTCEA-CT - Curitiba Airspace Control Detachment;

EPI - Instructor preparation traineeship;

EUROCONTROL - European Community to the International Convention for Cooperation for the Safety of

Air Navigation; Eurocontrol signatory countries: Albania, Germany, Former Yugoslav Republic of

Macedonia, Austria, Belgium, Bulgaria, Czech Republic, Cyprus, Croatia, Denmark, Slovak Republic,

Republic of Slovenia, Spain, Finland, France, Northern Ireland, Hellenic Republic, Hungary, Ireland, Italy,

Grand Duchy of Luxembourg, Malta, Moldavia, Principality of Monaco, Norway, Netherlands, Portugal,

Romania, Sweden, Switzerland and Turkey. ICA - Aeronautical Command Instruction;

ICEA - Institute for Airspace Control;

IN - INSTRUCTOR;

NOTAM - Notification to Airmen;

OI - Instruction Order;

PHO - Programme of Operational Qualification;

PUD - Didactic Unit Plan;

RICEA - Airspace Control Incidents/Accidents Report;

SCI - Fire Fighting Section;

SDOP - Operations Subdepartment;

SIATO - Operational Instruction Section;

SISCEAB - Brazilian Airspace Control System;

A system created to provide the Aeronautics Command with a structure capable of integrating the Bodies and Systems that participate in the control of National Air Circulation, within the limits of their respective attributions.

SRPV-SP - São Paulo Regional Flight Protection Service;

SSIATO-CT - DTCEA-CT Technical and Operational Instruction and Assessment Subsection;

TATIC - Total Air Traffic Information Control; and

TWR - Aerodrome Control Tower.

CHAPTER 1

1 INTRODUCTION

The issue of training for practical instruction in the training of professionals of SISCEAB's operational area, in recent years, has been adopting a more formal character. Besides, there is a greater concern in training the instructor, the main character of this process, who will also act as a planner of the instruction activity and not a mere reproducer of the planned action.

The instruction for practical activity has specific characteristics, different from a theoretical instruction. When the final activity of the subject to be trained involves risk, the instructor's formation inspires more care.

The focus of this research is the practical air traffic controller instructor who acts in instruction during actual operation and is responsible for the trainee during training and the final training of the control professional.

The practical instructor also acts in moments such as the controller's area or function reallocation. A controller who arrived transferred from another regional and needs to be trained for the new area to be controlled, or stopped acting as tower controller (TWR) and moved to an approach control centre (APP), even if these professionals already have years of operation. This indicates that, at all times, controllers, even experienced ones, need to be trained for the practical function they will perform and for such it is extremely important the performance of a practical instructor specially trained to train controllers.

1.1 Context, relevance and justification of the research work

We can notice the importance of the subject matter of this research when we find the subject in several documents of the International Civil Aviation Organization (ICAO), such as the ICAO Document 9868: Instruction (PANS) (2006) and Document 7192- AN/857: Training Manual (2004).

In the document addressing competencies for instruction RLA/06/901 - RCC/4 NI/03 (Regimento Aeronáutico Latino Americano 06/90; Regional Communication and Coordinations 4 NI/03) (2010) of ICAO, in its chapter 10, it stresses that it is important to consider aspects related to human factors and the planning of technical instruction in aviation. It points out that the high level of automation leads to the occurrence of several problems related to human factors. It highlights as a main objective to outline a uniform planning for the training of the air navigation professional.

The proposal is that this uniformity will occur by standardizing the practical-operational instructor training and the air traffic controller instruction, reducing the impacts mentioned in the document in this aviation sector in Brazil.

In addition to being present in documents, this issue is a topic of discussion at several regional and global ICAO meetings.

The training of the air traffic controller's practical-operational instructor is directly related to the quality of the ATC professional who will work in the System. By improving the quality of air traffic control services it can be inferred that there will be an increase in operational safety.

1.2 Scope limitation

The air traffic controller training is divided into three distinct but fundamental parts. The first part is theoretical instruction, fundamentals and essential knowledge for the subsequent stages and professional performance. The second part is simulated instruction, where theoretical knowledge is used in practice with total safety by means of simulators. The third part is practical instruction, the internship given in real time and which therefore requires a great deal of care.

This work has as its central focus the training of the instructor to act in the internship, that is, in the training of the actual practice of the air traffic controller.

It is common to find in some organizational cultures the concept that someone who performs a given function is perfectly capable of teaching it to another person, however this "transfer" of knowledge does not happen this way, it requires specific training for such, and the more complex the activity, the better the training of the instructor has to be.

This work comes as a response to this need presented by the Brazilian Airspace Control System, proposing a specific training for the qualification of the air traffic controller's operational practical instructor.

1.3 Research question and objective

This paper intends to present the national and international panorama of training and capacity building of practical-operational instructors of air traffic controllers (ATC) and, through analysis and studies of ATC instructor training and needs of the Brazilian Airspace Control System, propose an adequate training for these professionals in line with international guidelines.

According to the FAA's *Air Traffic Technical Training* 3120.4L, the operational practice instructor must fulfill theoretical and practical training, besides a supervised internship for instruction. This professional is submitted to a council that may or may not approve him/her for this type of instruction, depending on his/her performance during the training. Like the documentation cited above, several others developed by competent bodies in the international scenario, provide specific training for the practical-operational air traffic controller instructor or OJTI (on-the-job training instructor).

The document, in Brazil, that foresees the need for the training of the practical instructor of the CTA (air traffic controller) is the Aeronautical Command Instruction 100-18 Licenses and Certificates of Technical Qualification for Air Traffic Controllers in its item 8.5 which contains only that there shall be a trainer preparation stage and that this shall be planned and programmed by the Regional Organization of jurisdiction (8.5.1) and that this stage shall be applied and supervised by the ATC organ (8.5.2).

The product of this dissertation is a proposed standardized solution for CTA's practical instructor training. The suggested training content provides a solid knowledge base for practical instruction as well as for instructors to operationalise the OHP.

1.4 Methodological approach to research

For research data to be reliable it is necessary to work rigorously and use a specific method so that the data is valid and reliable.

In order to choose the most appropriate methodology it is important to know and delimit the problem, which in the case of this research is the training of the practical instructor of the air traffic controller.

After documentary research and questionnaire application to air traffic controllers and controllers' instructors who work in the System, it was noticed that, when the training for practical instruction happens, there is no standardization, each CINDACTA has a different training and it does not always contemplate the necessary information and knowledge for the practical instructor's performance. Although this training is foreseen in ICA 100-18, there is no "how to do it", the Aeronautical Command Instruction only says that it must exist but does not describe how it should be done.

The problem that is the object of this research, due to its complexity, is not only related to an area of expertise, a multidisciplinary vision is required to search for the solution of the problem encountered. As quoted by Laville and Dionne (2010) in their book "The Construction of Knowledge":

> "The real, it is thought, should be approached in its globality, as a system of interrelated factors." (LAVILLE AND DIONNE. 2010 p. 44)
>
> "What develops, then, is a multidisciplinary approach, which consists of approaching research problems by appealing to the various disciplines of the sciences that seem useful to us." (LAVILLE AND DIONNE. 2010 p. 44)

When thinking about the training of the controller's practical instructor, it is essential to think in areas that go beyond the operational issue. It is obvious that a good air traffic control instructor needs to be a good professional, with full domain of the theory and practice of the execution of his final activity. However, we cannot forget that, to be an instructor, he will also need pedagogical, psychological and even legal knowledge related to the performance of the activity. Regarding the classification of this work:

Table 1: Research classification.

Nature of the research	Applied
Form of approach to the problem	Qualitative
Purpose of the research	Exploratory and descriptive
Technical procedures used	<ul><li>Survey</li><li>Bibliographic research</li><li>Documentary research</li><li>Exposure-Fact Research</li><li>Participatory research</li></ul>
Scientific method	phenomenological

1.5 Sources of Information

The sources of information used for the development of this work were bibliographic research and documentary analysis through reports produced in investigations of aeronautical accidents and incidents. Finally, data were collected through questionnaires applied all over Brazil to professionals who perform air traffic control activities. Part of the sample also works as instructors in operational agencies and the others work only in air traffic control.

The data collection presented some problems as will be described in a later chapter, however they were fundamental for the development of this dissertation.

1.5 Training / capacity building

Since ancient times, training has played a fundamental role in the development of mankind. At each discovery or invention of mankind, the transmission of this knowledge, skill or technology becomes necessary for the progressive evolution of civilizations.

In the context of SISCEAB (Brazilian Airspace Control System), training becomes fundamental in the dissemination of theoretical technical knowledge and in the development of specific skills in the various areas of air navigation, seeking to train professionals in the performance of operational activities.

In this paper training and capacity building have the same meaning since the main objective is to make professionals able to perform the task of practical instructor through specific training for practical instruction.

The training also appears as a way to solve difficulties and possible problems that arise in the execution of activities necessary for the operation of the System. The role of training is always a resource to maximise results and achieve objectives as cited by Bomfin (2007).

There are several terminologies to define the word training according to the perspective of authors who study the subject, they are:

"Training is any activity that deliberately seeks to improve a person's ability to perform a task"

OATLEY; HAMBLIN, 1978 (p. 19)

"Training is a sequence of experiences or opportunities designed to modify behaviour to achieve a stated goal." HESSELING;

HAMBLIN, 1978 (p. 18)

"Training is the educational process, applied in a systemic way, through which people learn knowledge, attitudes and skills according to defined objectives." CHIAVENATO, 1985 (p. 288)

"Training in the company is the action of training and qualification of manpower, developed by the organization itself, in order to meet its needs." TOLEDO, 1986 (p.88)

"Training must take into account the whole human being, thinking, feeling and acting." KIRÁLI; CARDOSO, 1997 (p. 102)

"...get someone to be able to do something they have never done before, and do it without assistance from the one who teaches." CARVALHO; BOOG, 1999 (p. 127)

It will be shown in a later chapter, some pedagogical theories that complement the definitions described by the authors above and that are necessary in the training of the CTA practical instructor.

Next, it will be explained how the research was developed and the results found by it. In this way, a diagnosis of the practical instructor training of air traffic controllers in Brazil today will be outlined and the need to improve this training will be highlighted, due to the risks that the failure of this training represents to the System.

1.7 Structure of the work

Chapter 1 - The first chapter presents the work's introduction and provides a contextualization of the research's objective as well as shows its delimitation. It also shows the relevance of the subject and the need for the development of a specific study to address the problem presented, which points out the need to improve the professional skills of practical air traffic controller instruction. This chapter also presents the research classification, the sources of information and finally briefly addresses the subject of training and capacity building.

Chapter 2 - This chapter presents a diagnosis of the air traffic controller practical instructor training in Brazil. The information was collected from the analysis of incident and air accident reports filed at the ASEGCEA, from the application of questionnaires to air traffic controllers and controllers' trainee instructors and from the analysis of reports that record the creation and development of the only course for practical instruction existing in the SISCEAB.

<u>Chapter 3</u> - Chapter 3 describes the air traffic controller practical instructor training in the United States, EUROCONTROL (European Community to the International Convention for the Safety of Air Navigation Cooperation), England and Singapore. In the international context, the controller's practical instructor is called *"on-the-job training instructor"* (OJTI) and the training of such professional is standardized in order to make the training more efficient, seeking uniformity of procedures and instructional actions.

<u>Chapter 4</u> - Chapter 4 discusses the Operational Qualification Program - PHO. This program was developed by military personnel of the Brazilian Air Force belonging to SISCEAB and with extensive experience in air traffic control and practical instruction in the

area. The work describes how the operational stage should be structured and establishes standard procedures for the air traffic controller qualification process.

<u>Chapter 5-</u> In chapter 5, basic pedagogical knowledge is proposed and explained, which is the minimum for the practical air traffic controller instructor to be able to accomplish his task which is operational training.

<u>Chapter 6</u> - In chapter 6 the problem studied is addressed again which is the training of the practical instructor of air traffic controller and conclusions are presented.

CHAPTER 2

2 DIAGNOSIS OF AIR TRAFFIC CONTROLLER PRACTICAL INSTRUCTOR TRAINING IN BRAZIL

2.1 ASEGCEA survey conducted in August 2011

In March 2011, the researcher requested authorization from the competent authorities to have access to the incidents and air accidents reports filed at ASEGCEA with the purpose of continuing the research regarding the practical instructor's training of the air traffic controller. This information, which was extremely important, had the purpose to verify if, possible instructor's failures due to lack of adequate training during instruction moments could lead to the occurrence of aeronautical accidents or incidents.

The research conducted in the RICEA (Airspace Control Incidents/Accidents Report) took place in August 2011 at the ASEGCEA (Brazilian Airspace Control System Operational Security Advisory). The reports analyzed were the ones from 2008 and 2009 because the ones from 2010 and 2011 had not been carefully analyzed by the ASEGCEA team, the central body of SEGCEA, belonging to the DECEA (Department of Airspace Control) structure and directly linked to the DECEA Director General.

The ASEGCEA has the attribution of analyzing all documentation from the investigation of an Aeronautical Accident or Serious Aeronautical Incident, with airspace control involvement and Analyze, under the operational safety point of view, the Final Report of the post-accident Flight Inspection, related to an Aeronautical Accident or Serious Aeronautical Incident.

During the research carried out in the referred documents, data highlighted in relation to aspects related to failures in the air traffic controllers (ATC) training, which can be attributed to the ATC practical instructor's lack of proper training. Other serious failures which generated aeronautical accidents or incidents that occurred in the years 2008 and 2009 are directly related to the moment of the practical instruction and in the report itself is explicit the instructor's lack of instructional skills.

The determining or contributing factors that most drew attention in the reports of air traffic accidents and incidents were those directly related to instruction, that is, at the time of instruction and as a consequence of instruction, the controller's lack of skill and the non-application of knowledge of air traffic standards and rules by this professional.

As described in ICA 100-5, Air Traffic Incident Investigation, January 2003, the lack of ability presented by the controller is the situation in which an operational failure is associated to the air traffic controller's degree of skill in the execution of a procedure and/or in the application of a method during the provision of air traffic services. Thus, it is inferred that there was a failure in the training of the professional who presents such a deficiency.

With regard to knowledge of standards, ICA says that it is the situation where the operational failure is associated with the level of knowledge, by the controller, of the rules and/or procedures applicable by the ATS unit, during the provision of air traffic services. This is clearly a problem related to the cognitive domain, i.e. it concerns the lack of knowledge of the rules to perform the control activity. However, as this failure is repeated very often, and often by the same controller. Thus, we can deduce that the norms may be being ignored by the professional, which is no longer a lack of knowledge of the norm and becomes a norm infraction, which relates the problem directly to the work with affective domain of the professional when being trained, i.e., the controller is not attributing the necessary value to the performance of his activity, ignoring the consequences of his failures.

Another factor that was repeated in many risk situations was the non-use of standard phraseology, incurring in the use of colloquial language, which we can also attribute to the lack of work in the affective domain of the CTA in their training.

Problems were also found with traffic planning and coordination, which appeared as contributing factors in most of the accidents and incidents reported.

Thus, the reports at the end, as a recommendation, referred the controller to an operational recycling and it was not uncommon to find controllers who had already been through several recycling sessions because they had been involved in other accidents/incidents and in some cases for the same reason.

Today, few SISCEAB (Brazilian Airspace Control System) professionals have access to the instruction courses offered by the System, but it is noticeable the progressive concern of the authorities in solving this problem. Even in training courses for instruction, we find a predominance of the valorization of the cognitive domain to the detriment of the affective and psychomotor domains (motor skills), which we will address in detail in a later chapter. This way, for lack of knowledge about the subject, the instructors do not work their trainees in integrality, that is, taking in consideration that the student/trainee is not only intellectual but also needs involvement and respect for the professional activity as well as to develop motor skills for the performance of a given function, as it is the case of air traffic control.

The analysis of the RICEA presented difficulties in extracting the information necessary for the research, since the reports do not present a standard model for all the regional offices. Furthermore, there is no space with a summary of the information to facilitate the search. It was necessary to read the reports in full in order to analyse the data and verify what was or was not linked to the researched subject.

The reports, even for the seriousness they represent, were extensive and well detailed but it would be interesting to have a summary table with the contributing and determining factors of the reported accidents/incidents to facilitate data processing by the ASEGCEA team.

Another interesting aspect would be to have a standard report model for all the regionals. This would facilitate the search for information and the processing of the same with the aim of solving the possible problems presented by the System and/or professionals involved in the situation of danger/risk presented.

In tables 2 and 3 below are the data compiled from the research conducted in the RICEA of the years 2008 and 2009 of the four CINDACTA and SRPV-São Paulo. The percentages in the tables do not represent the sum for the total of reports (100%), because in one report more than one problem was presented. They are the part of the whole that demonstrated the deficits highlighted in the table.

Table 2: Airspace control incidents/accidents for the year 2008

2008				
	Total reports	Problems with CTA skills	Problems with standards and air traffic knowledge	Problems related to instruction
Operational centre A	64	37,5%	32,8%	20,3%
Operational centre B	18	38,9%	88,9%	11,1%
Operational centre C	13	46,1%	15,4%	0%
Operational centre D	10	90%	50%	10%

| Operational centre E | 19 | 42,1% | 57,9% | 10,5% |

Table 3: Airspace control incidents/accidents in 2009

2009	Total reports	Problems with CTA skills	Problems with standards and air traffic knowledge	Problems related to instruction
Operational centre A	68	50%	33,8%	23,5%
Operational centre B	20	45%	40%	10%
Operational centre C	16	50%	31,2%	6,25%
Operational centre D	11	63%	66,6%	18,18%
Operational centre E	21	33%	57,14%	9,52%

2.2 Application and results of the questionnaires

For this research, two questionnaires were prepared due to the fact that they were addressed to professionals with different activities. One of them was applied only to air traffic controllers and the other to controllers who work in the practical instruction of controllers.

This instrument was applied in the four CINDACTA and had the participation of some controllers and/or instructors of the SRPV-SP.

In some cases, the questionnaire was applied to students of course CTP006 (SISCEAB Practical-Operational Instructor Course) during the period in which they were taking the course due to the fact that the researcher is an instructor and coordinator of the referred course.

During the analysis of the questionnaires, some data called our attention, such as some regional centers that presented a significant gap in relation to the others. The operational centre "D" showed a greater lack of knowledge related to instruction, especially practical instruction. Some fundamental pedagogic concepts, often intuitive for those who work with instruction without proper training, were misunderstood. Almost all the practical instructors did not have any instruction course, although they agreed it was fundamental to the performance of the instructor function.

The most striking data extracted from the questionnaire applied to air traffic controllers are shown in the tables below:

Table 4: Choice of Instructors

Choice of the instructors to teach the operational training course				
The oldest	The most modern	Military with more experience in the field	Someone who has specific training for operational instruction	Don't know
23%	0%	47%	24%	6%

Source: standard survey questionnaire

This table shows that 70% of the practical instructors in the System, who were part of the sample, do not have training for practical instruction; only 24% had the EPI or CTP006.

It is important to remember that CTP006 is not specific for the practical instruction of air traffic controllers, although it offers an excellent theoretical basis for general practical instruction. With regard to PPE, as seen earlier, it is not standardized, leaving this stage open and being applied in very different ways in each region.

Table 5: Interference of instructor training with CTA training

Interference of the operational internship instructor's lack of training in the technical-professional training of the student trainee	
YES	NO
94%	6%

Source: standard survey questionnaire

What we can extract from the data above is that the controllers perceive the direct relationship that exists in the adequate preparation of the practical instructor for the training of the professional who will work in air traffic control.

Table 6: Failures in instruction X operational safety

Faulty instruction generates risks to operational safety

YES	NO
100%	0%
YES	NO
100%	0%

Source: standard survey questionnaire

The information in the table above is very significant for the research, since 100% of the controllers who answered the questionnaire state that failures during instruction generate risk to operational safety. This data can be confirmed if we cross-reference the information obtained in the RICEA survey, where a large number of reports from 2008 and 2009, point out as a contributing or determining factor for the accidents/aircraft incidents highlighted, the instructor's failure during the instruction.

Below are spontaneous records transcribed from the questionnaires with the same words used by the controller who wrote them.

"Sometimes some operators of operational bodies are forced to do instructional course without having any vocation."

"I think that the idea of turning a controller into an instructor in SISCEAB is not taken very seriously. For, where I come from, regardless of the profile of the controller, with two years of service the operator becomes an instructor."

"Our biggest problem, regarding instruction, is due to the lack of CTA instructors prepared for the type of instruction (practical). I had the opportunity to work with instructors who had less than one operational year."

"All instructors in SISCEAB, without exception, must have courses of instruction"

"I believe SISCEAB should look more at its instructors, as most of them do not have the vocation or instruction to act as such."

"No one in SISCEAB wants to be an instructor by choice or vocation... the fear of getting involved in air traffic incidents/accidents linked to coordination (in addition to the commitment to my duties) that made me decide to become an instructor when it was proposed to me, even with little time in the field and no prior preparation."

"Some people in SISCEAB are approved early for instruction and, in addition, they have no course of instruction."

"There is no adequate training for instructors nor standardization of instruction. So, each instructor passes in the internship what he thinks is relevant according to his experience, which doesn't always match reality. Some instructors don't really want to be in this field.

"Even though the choice of instructors is by seniority, the quality of instructors is decreasing as most of those who are "old" nowadays have no more than 4 years of operation."

The data presented below is pertinent to the questionnaires applied to the practical instructors of air traffic controllers.

Table 7: Instructors' instruction time

Time instructors deliver instruction			
1 year or less	between 1 and 5 years	between 5 and 10 years	+ over 10 years
20%	16%	32%	28%

Source: standard survey questionnaire

The table above makes clear that 68% of the instructors who answered the questionnaire have less than 10 years of experience in the area, and 20% have not even been acting as instructors for one year.

Table 8: Instructors with training course

Instructors with training course for trainee classes offered by the system (CTP006 or EPI)	
YES	NO
25%	75%

Source: standard survey questionnaire

A factor that drew a lot of attention was the fact that, among those who answered the questionnaire, 75% work in practical instruction and have no preparation for the performance of the function, against only 25% who had the opportunity to take EPI or CTP006, which are the only courses/internship offered by the System. It is important to stress that the internship and course mentioned above do not fully meet the training needs for the development of competencies to work in practical air traffic control instruction.

Below are spontaneous records transcribed from the questionnaires with the same words used by the instructor who wrote them.

"My internship was bad and I don't blame my peers for my unpreparedness."

"...instructors are not chosen...if there is not enough instructor, the most senior military (in today's reality, the least modern) are called in to make up the team of instructors, regardless of the military's teaching ability."

"Local PPE (DTCEAs) are often applied by people without specific technical knowledge for instruction."

EPI - Practical Instruction Internship - provided for in ICA 100-18

"My placements were almost informal, with no briefing or debriefing, done by instructors with no specialist training, just passing on the experience,"

"The lack of standardisation and training of instructors increases the risk of the operation, as they create their own ways of controlling that most of the times violate the technical air traffic norms and give amateurism in a profession with high technical criteria."

2.3 Analysis of the reports of the CTP006 - Practical-Operational Instructor Course of the SISCEAB

In the Brazilian Airspace Control System (SISCEAB), most of the professional training processes include practical instruction activities, conducted by instructors who do not have the necessary training to perform such activities. As far as instructor training is concerned, in 2007 the System had only one (1) course, the Standardization Course for Instructors (CTP001), which was entirely directed to train instructors to teach theoretical classes to collective classes, emphasizing the expositive technique and the use of the platform. Despite the perceived difficulties, however, the practical training for the various SISCEAB professionals took place.

The ICEA (Airspace Control Institute) together with DECEA (Airspace Control Department) through the DCT-P (Professional Training and Capacity Division), activated a Working Group that met in the week of July 31 to September 03, 2007, at CINDACTA III in Recife. The objective of this group was to elaborate a training course for the practical instructors of the several technical areas of SISCEAB (Brazilian Airspace Control System).

The team was composed of 8 (eight) military personnel being them:

- 1° Lt Osmário Guedes dos Anjos, air traffic control specialist
- 1° Lt Harrison Bezerra Rodrigues, meteorologist
- 1° Lt Antonio Robson de Carvalho, air traffic controller
- 1° Lt Jorge Augusto Martins, air traffic controller also graduated in pedagogy,
- 2° Ten Tereza Cristina Buonocore Nunes, pedagogue of the Complementary Group
- SO Cláudio Luiz Guanabara, graduated in psychology
- 1S Cícero Morais Filho, meteorologist

- 1S José Coelho Rodrigues, meteorologist with a degree in geography.

The work started with the *brainstorming* technique from which the fundamental topics for the composition of the pedagogical documentation of the course, such as the Minimum Curriculum - CM and the Didactic Unit Plan - PUD, were extracted. The disciplines chosen by consensus were: Didactics for Operational Instruction and Psychological Aspects in Instruction. The first one contemplated planning, teaching objectives, practical instruction and evaluation; all didactic units of the discipline were built with a specific focus on operational practice. The second approached interpersonal relationships and psychological and cognitive factors that influence practical instruction. The subjects were duly broken down and synthesized into verifiable objectives, following Bloom's Taxionomy of Educational Objectives (1972) and Elizabeth Simpson's Psychomotor domain (1972) to enable a proper evaluation of their achievement. These objectives were discussed by the various professionals present so that they could be adapted to the needs of the system.

Due to the need to adapt the course to the workload equivalent to 5 (five) working days, it was first established as a prerequisite that the students had the Instructor Preparation Course (CPI), or the SISCEAB Instructor Standardization Course (CTP001). This pre-requisite would partially guarantee to the students the minimum necessary conditions for the adequate follow-up of the new course contents. This pre-requisite was later waived when the course was extended to two weeks, for a total of 10 (ten) working days of classroom instruction, and also to include more basic contents, which were included in the CTP001 course.

During the meeting, the Weekly Work Program (WWP) was also elaborated, with the forecast of the distribution of the subjects to be approached in the instruction in increasing order of content complexity along the instruction days, the Global Evaluation Chart (GEM) which foresees the evaluations to be done to verify the reach of the class objectives and their respective weights, the Cover Sheet which determines the target public besides foreseeing the pre-requirements to the course accomplishment and its general objective. Besides, it was made a list of possible instructors qualified to teach the course, which was named CTP006, Operational Practical Instructor Course.

At the end of the meeting some actions were recommended to proceed with the elaboration and implementation of the course:

- That the specifications recorded in the PUD be observed for the purpose of implementing the course (desired requirements of instructors, teaching resources to be employed, etc.);

- That the DECEA Sub-Department of Operations (SDOP), together with its Divisions, begin to study the proposal of training its practical instructors, starting with the CTP006 course, which could be followed by other stages, such as the Instructor Standardization Internship (EPI), recommended within the Air Traffic Control scope through ICA 100-18;

- The DECEA Sub-Department of Operations (SDOP) will also study the proposal to create evaluation forms for trainees in the various Airspace Control specialties, as well as the establishment of more precise criteria for the qualification of its professionals. This action was carried out with CIRCEA 100-51 and CIRCEA 100-52 (although the proposed forms are not yet considered ideal);

- That the 2^a. Phase of the activities of the working group, with the purpose of preparing the teaching resources to be used in the course and preparing the corresponding evaluation instruments, given the peculiarities of this course and the nature of the subjects it covers, which were unprecedented in the System at that time;

- That the initial application of this course be scheduled for validation, in order to make the necessary adjustments before its availability to the System. For this application, it was recommended the presence of the members of this WG as teachers and students previously selected by this team, in order to make such adjustments effectively.

From 7 to 11 October 2007, also at CINDACTA III, the WG met again to continue the work. These were started from the reorientation of those present in relation to the desired parameters for the production of teaching material, as well as the guidelines for the development of test items, since it was established for this course the final application of an assessment based on multiple choice questions. The more thorough analysis of the detailing of the material led the G.T. to the reassessment of some aspects of the initial proposal of the course, such as:

- name of the course, since the G.T. proposed that the course should receive the name of "Basic Training Course of Practical Operational Instruction", since its proposal is to "provide to the students a minimum theoretical learning necessary for the exercise of practical operational instruction in the SISCEAB". In this way, the course constitutes the beginning of the instructor's training, understood as a process capable of filling a gap until then existing in the System. The conclusion of the training would only occur with the other operational and regulatory requirements, established for each specialty;

- adjustments to the Teaching Unit Plan, the Cover Sheet and the Weekly Work Programme:

Other recommendations were established at that second meeting:

- That the last necessary adjustments were made before the application of the course in the System. It was intended to make the application itself, as continuity of the instructors' group activities and, immediately after the course, to make the debriefing of the activities with all the team present, adopting, from there, the corrective measures;

- Train multipliers (instructors) for this course, based on the candidates' profile, appreciation by the G.T. and their training by the G.T;

- [a]As it was proposed in the report of the 1st Phase of activities of the G.T., that this course would not replace other training processes of instructors that exist today in SISCEAB, but would only complement them.

According to the schedule drawn up by the WG at its last meeting in October 2007, an experimental validation course was held from 12 to 16 May 2008. The body of instructors for this course was formed by the WG members themselves so that a comparison could be made between the activities envisaged in the planning work and the results obtained.

As expected, the class was composed of elements with various specialties of SISCEAB, since the objective is to enable the instructor to apply a program of practical operational instruction within any area.

Within the weekly work program prepared by the WG (PTS), the activities were initiated in the second half, right after the general briefing of the ICEA to the students. In a second moment, a group work was carried out where the expectations of the students in relation to the course were sought. In this particular, mistaken expectations were perceived in relation to the training of the students of that course, as they imagined that they would leave the ICEA with an instructional program ready to be applied in each area of their department and, where in fact, the intention was to prepare them to serve as facilitators in the teaching-learning process, i.e., builders and applicators of a specific program for each SIATO (Section of Technical-Operational Instruction) of their respective Unit.

In search of data for the validation of the course, a thorough daily confection of the students' criticisms and suggestions for modifying the course was requested. The unanimous opinion was the issue of the reduced workload for the application of all the content, making discussions related to them impossible. In all the days, there was the need to go beyond the timetable to compensate, at least partially, the little time foreseen to teach the contents considered minimum, since the WG, to attend the request of accomplishing the course in one week, withdrew information that, later, in council, were considered fundamental for the practical instructor's qualification. Besides, there was questioning and participation of the students in search of knowledge they considered indispensable to their performance as instructors.

It was verified the need to increase the practical work within the course as the implementation of moments of construction of didactic materials for instruction such as lesson planning, evaluation sheets and others. It was also perceived the need to include moments of discussion in order for students to assimilate the theoretical content that underpins the practice.

The confirmation of the above notes was given by the poor performance of students in exercises applied in the classroom with the aim of checking whether they could make the transfer from theory to practice through videos that simulated the operational reality.

Throughout these years, the application of the CTP006 course has generated satisfactory results, especially in CINDACTA 2, where it has been formalized the requirement that all ATC practical operational instructor candidates must first take this course, since there is no specific course for practical instruction focused on air traffic control. That Center has resized its PPE (Practical Internship for Instruction) according to this new system.

From the contents taught in CTP006, some CINDACTA 2 professionals with extensive experience in air traffic control and with years of instruction in operational training, organized themselves to develop the PHO (Program of Operational Qualification), with the objective of standardizing the practical instruction. This program has been successfully applied in CINDACTA 2 by the professionals who developed it and it is being progressively implemented in the other regional airports. The PHO will be detailed in a later chapter.

2.4 Brazil and the world scenario

After knowing the air traffic control instructor national scenario, it is important to contextualize it in the international scenario. In the following chapter, we will point out the regulatory agencies responsible for this instructor's qualification, the documentation which support this formation, how the trainings are given, their duration, the instructor's certification and other fundamental aspects related to the subject and which are extremely important for this research.

CHAPTER 3

3 *ON-THE-JOB TRAININGINSTRUCTORS* IN THE STATES

UNITED STATES, EUROCONTROL, ENGLAND AND SINGAPORE

3.1 The Study

The purpose of this chapter is to present a documental and bibliographic study of the training of practical-operational instructors in Air Traffic Control (ATC) in the United States of America, EUROCONTROL, England and Singapore. These professionals are known worldwide as OJTI *(On-the-Job Training Instructors)*.

The instruction activity performed by OJTI/CTA takes place with the trainee who has completed his theoretical training and simulated practice, and who is assigned to an air traffic control agency to exercise the activity for which he has been prepared to perform. This type of training is of fundamental importance in the case of apprentices of the air traffic control function, since a determined period of time is required, with adequate guidance, so that the student may become a professional of the area and become familiar with the real practice and the peculiarities of the operational body in which he/she is inserted, thus acquiring mastery of the functional activities. This way, the practical-operational instructor, or OJTI, responsible for the trainee's learning period in a real situation, is a fundamental figure. This way, the operational safety of the execution of the task is preserved, when the trainee is still in the professional training phase.

To play a role of such importance, pedagogical, organisational, technical knowledge specific to the CTA's area of activity and content from the area of psychology are required due to the complexity of the instruction activity in a real situation.

In this work, it will be briefly described some trainings of practical-operational instructors in the above mentioned countries, exemplifying the contents given as well as the time of the course or internship.

Another important factor to be addressed is the form of qualification of these professionals who work with instruction in actual practice with air traffic controllers and the processes required to maintain this qualification, which is fundamental for the preservation of the quality of instruction and for operational safety.

3.2 History of OJTI capacity building

In June 1990, a forum examining various aspects of *On-the-Job Training* (OJT) was held at EUROCONTROL's *Institute of Air Navigation Services* (IANS). The training of the On-the-Job Training *Instructor* (OJTI) was discussed and, based on the results of the expert analysis, the development of specific training for this professional was proposed.

The first OJTI training programme was delivered at IANS in December 1994. This course was developed as a result of work conducted by EUROCONTROL training specialists from France, Germany, Ireland, Switzerland and the UK.

In 1998 a team composed of specialists in the area of training was created, known as the *Training Focus Group,* which meets annually for the purpose of monitoring the OJTI courses delivered, reviewing the content and proposing changes to the course plan.

Over time, this group of experts also became responsible for monitoring, adapting, planning and evaluating the periodic OJTI updates provided for in regulations in the form of a refresher course.

In this way EUROCONTROL maintains a standardization of the instruction provided by the courses related to OJTI training and also preserves their quality through evaluations made systematically with the aim of safeguarding operational safety.

3.3 OJTI capacity building governing documents

Countries that provide for the training of practical-operational instructors (OJTI) have documents that regulate the training of such professionals, establishing minimum standards and requirements to be met so that the training can take place effectively and safely.

Below, there is a table containing the documents and the issuing and controlling bodies of the professional training mentioned.

Table 9: Controlling bodies for capacity building

1. Country/Agency	2. Document	3. Regulatory body

Country	Document	Responsible Body
England	CAP 744, *UK Manual of Personnel Licensing - Air Traffic Controllers in Accordance with EUROCONTROL Safety and Regulatory Requirement number 5 (ESARR5)*	CAA *(Civil Aviation Authority)*
Singapore	*Manual of Standards - Licensing of Air Traffic Control Personnel* (version 1.1 of 06/05/2010)	CAAS *(Civil Aviation Authority Singapore)*
EUROCONTROL	*Air Traffic Controller Training at Operational Units* HUM.ET1.ST05.4000-GUI-01 and ATCO *development Training: OJTI course* HRS/TSP- 004-GUI-06 (08.04.2004)	EUROCONTROL
United States of America	*Air Traffic Technical Training* (22 June 2005) 3120.4L, Air Traffic Technical Training	FAA *(Federal Aviation Administration)*

Source: CAP 744, ESARR5, Manual of Standards - Licensing of Air Traffic Control Personnel 2010; Air Traffic Controller Training at Operational Units 2004, ATCO development Training: OJTI course 2004, Air Traffic Technical Training 2005.

These documents foresee important factors for the success of this professional training, which will be better explained in the following topics.

3.3 Minimum requirements to act as OJTI

The responsible bodies, through specific required documentation, establish minimum requirements so that the CTA (Air Traffic Controller) professional may act as a practical instructor.

In England, the CAA (2009, p.40), in document CAP 744 - *Manual of Personnel Licensing - Air traffic controllers,* provides that in order to exercise the activity of OJTI, the applicant must:

- Hold a valid licence to act in air traffic control;

- Have successfully completed an OJTI course recognised and accepted by the CAA;

- Have at least two years of operational experience related to the discipline in which they intend to provide instruction;

- Have a positive evaluation for six months in the sector and operational position in which you intend to provide instruction.

The CAAS (2010), in the document MOS - *Manual of Standards - Licensing of Air Traffic Control Personnel*, determines the minimum requirements for acting as OJTI, establishing in chapter 3 the minimum experience and qualification requirements. Thus, in this manual it is provided that an OJTI must meet the following minimum requirements:

- Have a valid ATC licence;

- Possess positive evaluation in relation to the content in which he/she intends to provide instruction;

- Have passed the OJTI course recognised by CAAS;

- A minimum of two years' experience is required for the ATC to master more complex levels of air traffic control performance.

EUROCONTROL (2002) published ESARR 5 -Safety Regulatory Requirement, which states that the Airspace Control Department of each country belonging to EUROCONTROL shall provide authorisation to carry out the activity of operational training for air traffic controller (OJTI) only to candidates who can guarantee it:

- Minimum of two years experience, minimum level required to act as an instructor so that he/she can master more complex actions aimed at the operation contributing to the OJTI activity;

- Minimum of six months' experience in the discipline relating to the specific sector and operational position in which instruction is to be given;

- Have completed an OJTI course offered by EUROCONTROL;

- Have been approved in an operational council for the instruction activity.

ESARR 5 also requires the Airspace Control Department (DECEA) of each country to ensure that the controller providing training is certified by the competent authority and has a valid license to act as an OJTI. The said authority must also inform when the validity of the certification is expired or when it is identified that the CTA is not competent to conduct operational instructions.

OJTI should demonstrate performance in:

- Good interpersonal relationship;

- Communication skills;

- Motivational and general attitude to work more specifically to perform in instruction;

- Possess professional credibility among their peers;

- Must be considered operationally competent in acting as CTA.

The document further provides that OJTI should undergo a probationary period during which, from time to time, an assessment of instructional practices will be conducted by the sector responsible for the training.

It is recommended that the OJTI does not leave his/her operational functions so that he/she may continue to perform air traffic control activities with the skill required to perform the function. In case this does not occur, after two years, 25% of his/her working time must be assigned to operational activities, so that the referred professional may recover his/her skills as an ATC. The period dedicated to this workload distribution will depend on the professional, as this determination aims at ensuring that he/she obtains the necessary skill in the execution of operational activities.

The air traffic controller must always be prepared to return to his/her operational functions, and the aforementioned determination also serves to ensure that this professional keeps his/her technical skills up to date.

In the United States of America, the FAA (2005, p.45) has published the *Air Traffic Technical Training* manual *and in* section 2 called *On-the-Job Training* and *Position* Certification, states that to be eligible as OJTI the candidate must meet the following minimum qualification criteria:

- Be certified as a CPC *(Certified Professional Controller)* or FPL *(Full Performance Level)*;

- Be certified for at least six months in the positions in which they intend to provide instruction. If the applicant has come from another unit with OJTI experience in the

same position intended, the said professional may request his/her leave to instruct after 60h of internship.

- Be acting as CTA in the positions you intend to instruct;

- Be recommended by the immediate supervisor

Analyzing the minimum requirements demanded by the countries/agencies in relation to OJTI, it is clear the similarity and repetition of topics. This is due to the cohesion of the involved agencies, which aim at preserving safety in the performance of the practical instruction activity due to the fact that it takes place in real time. Not paying attention to the issues highlighted would represent a risk to operational safety.

3.5 OJTI training content

3.5.1 England - CAA

In CAP 744 (CAA; 2009, p.16), air traffic control training is defined as all theoretical courses, practical exercises including simulation and OJT, which offer the student the possibility to acquire and maintain technical professional skills that enable the provision of safe and high quality air traffic control services.

According to *EU directive* 2006/23/EC quoted in CAP 744 (CAA, 2009), the syllabus of the course delivered by CWMBran Training College offers three OJTI training courses, for CAA, for ICAO *(International Civil Aviation Organization)* and also for acting in instruction to attend military operations. These courses have the same syllabus, but cater for the groups separately due to the content of the discussions being different depending on the area of operation.

The theoretical subjects are:

- Teaching techniques;

- Human factors;

- Learning approach;

- Training planning;

- Preparation of an evaluation report.

Practical classes work:

- *Briefing* techniques;

- Techniques for monitoring learning development;

- Evaluation in operational practice;

- *Debriefing* techniques.

3.5.2 Singapore - CAAS

The reference documenting the content of the training is the *Manual of Standards - Licensing of Air Traffic Control personnel* (CASS, 2010). This manual advocates that for the OJTI course to be appropriate, it should cover specific content from different areas. Below is the proposed document divided into themes and topics.

Organisation of training:

- ATCO training content and objectives and training units;

- Planning - training unit plan;

- Structure of the training unit;

- Regulation.

Human factors:

- Teaching-learning;

- Teams and team interaction;

- Communication;

- OJTI profile and role;

- *Stress.*

Teaching techniques:

- *Briefing;*

- Demonstration;

- Communication;

- Follow-up;

- *Debriefing.*

Assessment and reporting methods:

- Assessment;

- Preparation of a report.

For Capelle Consulting (2010), an institution that provides OJTI training, the objectives to be achieved by OJTI course students are:

General objective:

- Systematically train a trainee in a work environment.

Specific objectives:

- Explain the importance of OJT in developing the productivity of the organisation;

- Describe the characteristics of the OJT instruction;

- List the benefits of OJT;

- Establish the four components of the OJT approach;

- Produce follow-up reports;

- Follow up OJT instruction with task analysis document prepared;

- Establish the training steps;

- Evaluate the trainee's learning;

- Establish the limitations of the OJT;

- OJT Resources.

The requirements for passing the OJTI course are to complete the course successfully and obtain 75% attendance.

3.5.3 EUROCONTROL

EUROCONTROL has published the governing document on *Air Traffic Controller Training at Operational Units* and the *ATCO Development Training: OJTI Course* (EUROCONTROL, 2004).

The OJTI Course recommended by EUROCONTROL is divided into five disciplines or modules.

Objectives and contents of the subjects:

Introduction:

- Purpose: To explain the structure, content and review process of the OJTI training programme.

This part of the course aims to prepare students for the activities they will develop throughout their training as OJTI.

Organisation of Training:

- Aim: To analyse the impacts of regulation on ATC training.

This subject covers the definition of basic training, appropriate level of training, Training units, Unit planning, structure, content, evaluation and application of planning, it also covers teaching techniques appropriate to OJT.

Other issues addressed are OJTI rules and responsibilities, communication channels available between instructor/instructor, retrieval activities for learning and performance problems, and support for student personal problems.

Aspects related to the regulations concerning the regulatory body such as competencies, responsibilities of the regulatory body and the school providing the training and of the individuals involved in OJT instruction (supervisor, controller, OJTI, Student and assessor) are also addressed.

Human Factors:

- Aim: to describe the human factors aspects related to learning, teamwork, communication and *stress* that impact on ATC training.

This subject covers motivation, behavioural theories of learning, *debriefing,* critique in training and learning theories. It also presents the importance of teaching objectives, writing teaching objectives, domains and levels of learning, teamwork, team interaction, mental models, behaviour and team cohesion.

Due to the great importance of the subject, verbal and non-verbal communication, communication barriers, the influence of communication styles on learning, conflict resolution strategies and evaluation are also addressed.

They also deal with the profile and role of OJTI, meaning, effect and symptoms of *stress* and stressor factors.

Training techniques:

- Objective: To apply appropriate teaching techniques to OJT training such as *briefing,* demonstration, communication, follow-up and *debriefing* based on the learned theory

This subject addresses the need for and preparation of the *briefing,* OJTI *checklist,* stages for structuring the *briefing,* need for exchange of information between instructors regarding the instructor and application of the *briefing* technique. In addition, it provides for the approach of the demonstration technique, application of the communication appropriate to operational instruction and monitoring of student learning. Regarding this last item, the subject provides not only the description and effectiveness of learning follow-up, but also the OJTI responsibilities, the support to be given by the instructor, the importance of records and record techniques.

Other issues dealt with are the techniques of questioning, category and error correction, forms, reasons, timing and control of intervention in *debriefing*, need, time, place, preparation, stages and application of *debriefing* techniques.

Evaluation and reporting method:

• Purpose: To explain the purpose of evaluation in training and the importance of proper report writing that represents the learning moment.

This subject addresses the need for assessment, the process within the organisation, factors interfering with assessment, OJTI's responsibility in this process and assessment techniques.

It also deals with the need, quality and use of reporting as an important tool for evaluation.

3.5.4 United States - FAA

The FAA (2005) provides the *Air Traffic Technical Training* as the governing document for OJTI training. In it is provided the course outline that follows below.

Learning process:

• Interpersonal relationships;

• Learning theories;

• Teaching techniques;

• Instructional resources.

OJT process:

• Instructional monitoring strategies;

• *Feedback - evaluation/debriefing;*

• Production of learning development reports;

• Motivation;

- *Coaching;*

- Types of evaluation for promoting OJT performance improvement;

- Course planning and evaluation.

OJT environment:

- Trainee's beliefs and preconceptions and their impact on training (Pygmalion effect);

- Qualities of an OJTI.

Team training process:

- Team skills and group interaction;

- Team training;

- Elaboration of team training plans.

Training documentation:

- Instructional documentation;

- Instruction records and common problems;

- OJT process registration.

3.6 Duration of OJTI courses

Below is a table presenting by country the duration of the courses available for OJTI training.

There are some differences in the duration of the course between the schools that offer OJTI training. Some offer a two-week face-to-face training course and others provide a distance learning phase and a one-week face-to-face practice phase, although EUROCONTROL recommends that OJTI training should preferably be conducted face-to-face due to the complexity of the contents and the activity the instructor will perform.

Table 10: duration of courses

1. Country/Agency	2. Duration Presential	3. Distance Duration	4. Source/ Schools
England	2 weeks	Allows 1 week distance if necessary	CWMBran and ASTAC
Singapore	5 days	-	www.saa.com.sg

EUROCONTROL	2 weeks	Allows 1 week's stay if necessary but advises against it	HUM.ET1.ST05.400 0-GUI-01 and HRS/TSP-004-GUI- 06
United States - FAA	3 days to two weeks	It admits part of the content at a distance when the course has more than three days	3120.4L and FAA55049

Source: HUM.ET1.ST05.4000-GUI-01; HRS/TSP-004-GUI-06; 3120.4L and FAA55049

The maximum class size for this course is three to eighteen students, with a maximum of twelve students allowed for EUROCONTROL, CAA England and CAAS Singapore.

3.6.1 EUROCONTROL

It is recommended that the OJTI training programme should be two weeks in duration, however, if there is any impediment to these two weeks being conducted face-to-face, one week can take place using distance learning.

It is recommended that the training be given in a fortnight, as it ensures a gradual development of the participant towards the required OJTI objectives, allows for the provision of remedial training should the need arise and provides the opportunity for consolidation of the knowledge covered throughout the training.

Note: Distance training is considered too difficult to manage for learners to achieve the objectives at the appropriate level and with the appropriate quality set by international standards.

3.6.2 United States - FAA

In the United States there is a range of courses offered by public schools and their duration varies from three days to two weeks in person, with or without distance learning.

The official FAA course has three days of face-to-face instruction only. FAA55049 (OJTI training) is intended to train OJTIs in the OJT environment and assist in meeting the recommendations of FAA3120.4.

The FAA provides for regular refresher courses every two years. Such training is offered in a three-hour course aimed at updating OJTI on learning-related aspects.

3.7 Controlling body of the OJTI activity

- England - CAA

In the UK, the CAA regulates air traffic services including personnel licensing, delegated authority under the *Civil Aviation Act* and in accordance with the articles of the *Air* Navigation *Order*. The Air Traffic Standards Division (ATSD) of the *Safety Regulation Group* is the authority responsible for regulating the licensing process, as per CAP 744 - *UK Manual of Personnel Lic*ensing (CAA, 2009, p.11)

- **Singapore - CAAS**

To meet the ICAO requirement, ATC must be authorised by the Aerodrome Navigation *Services Regulation Division (*AAR) to provide instruction in the operational environment if it meets the minimum requirements.

It is the responsibility of the ANSP (*Air Navigation Service Provider*) to ensure that these requirements are in line with OJT functions for air traffic instruction.

- **EUROCONTROL**

The regulation is done through the responsible expert group.

- **United States - FAA**

The FAA itself is the controlling body, according to *Air Traffic Technical Training* (FAA, 2005, p.13).

3.8 OJTI Certification

- **England - CAA**

CAP 744 (CAA, 2009, p.89) provides for OJTI certification through application for such a licence. In that part the licence for practical instructor assessor is also provided for.

Applicants for an examiner or OJTI licence must meet the requirements set out in part 2 of CAP 744: paragraph 5.2 for examiners and 4.1 for OJTIs.

- Possession of a valid licence;

- Have successfully completed an OJTI course recognised and accepted by the CAA;

- Have at least two years operational experience;

- Have a positive evaluation for six months in the sector and operational position in which you intend to provide instruction.

After successful completion of the course and meeting the prescribed requirements, the candidate may apply for their OJTI licence through a specific form provided by the CAA.

- **Singapore - CAAS**

In accordance with ICAO Annex 1 Chapter 4, the certifying authority is required to authorize the ATC to instruct as OJTI in an operational environment. Therefore the professional may only exercise the activity if approved by the CAAS. The requirements are the same as the ones foreseen by the CAA.

- **EUROCONTROL**

The conditions for the applicant to apply for their licence to act as OJTI at EUROCONTROL (2004) are:

- Have been an air traffic controller for at least two years;

- Demonstrate performance appropriate to the role;

- Be proposed or designated (subject to a general agreement from the team of controllers);

- Passed theoretical OJTI training in person;

- Passed practical OJTI training;

- Complete the probationary period assessed by the training unit;

- If approved you receive certification - a driving licence.

- **United States - FAA**

A board consisting of at least two persons must recommend the OJTI candidate based on personal attributes such as:

- Good interpersonal relationship;

- Communication skills;

- Motivation and attitude;

- Objectivity;

- Professional credibility.

This recommendation should go to the *Air Traffic Management* (ATM) team for final approval.

For the candidate to assume OJTI duties he/she must have successfully completed the FAA approved OJTI course. Completion of the course must be documented on the appropriate FAA form (3120-1, section VII), have approval for OJTI certification given by the candidate's supervisor who must make assessments and observations during the first OJT section. Documentation of this certification must be made on the proper FAA form 3120-1, section III.

The OJTI supervisor shall evaluate the candidate while the candidate is performing instructional activities. This evaluation shall occur no later than thirty days after assignment as an instructor and every six months (3120-1, section VI).

3.9 OJTI rules and responsibilities

- **England - CAA**

OJTI should be responsible for:

- The safety of the ATC service in relation to the service provided by the trainee at the time of instruction;
- Ensure that the ATC trainee:
    - Meet the necessary prerequisites for the internship - in the UK there is a licence to act as a trainee;
    - Have successfully completed the initial training course and be able to act in the position in which they will have practical instruction;
    - Have a medical certificate of the appropriate class;
- Determine and report on training progress through periodic assessments;
- Identify skills and possible deficiencies by providing remedial training;
- Assess the trainee to establish the level of competence they are at to recommend them as fit for role;
- Supervise controllers who have had their licences suspended;
- Monitors and reviews the instructional plan and proposes changes to it.

- Singapore - CAAS

Essentially OJTI's responsibility is:

- Preserve the safety of the ATC service during instruction;

- To ensure that the trainee acquires competence in the use of the new standards, procedures, techniques and equipment identified as essential for the execution of the task;

- Evaluate the progress of the training;

- Identify any knowledge or skill deficiency and recommend remedial training;

- Assess the ATC trainee as having or not having an appropriate level of competence to perform the activity;

- Supervise ATCs that have obtained negative evaluations;

- Review, monitor and propose changes to the training.

- **EUROCONTROL**

OJTI should be responsible for:

- Provide training in operational situations and may also act in the theoretical face-to-face phase of the OJT;

- Ensure operational safety during instruction.

- **United States - FAA**

OJTI should be responsible for:

- Use methods and techniques appropriate to the development of OJT instruction;

- Be familiar with the training process that will enable the ATC professional's certification;

- Provide OJT instruction;

- Preserve operational safety during instruction;

- Take responsibility for the facts occurred during the instruction.

3.10 Summary of capacity building

Cross-referencing the information between the training courses studied, below are the contents presented:

> Teaching techniques for CTA practical instruction;

> Human factors;

> Learning approach;

> Training planning (theoretical and practical classes);

> Fundamentals of evaluation;

> Preparation of evaluation report (theoretical and practical class);

> Evaluation in operational practice (theoretical and practical class);

> *Briefing* techniques (theory and practice classes);

> *Debriefing* techniques (theoretical and practical class);

> Techniques for monitoring and developing learning;

> CTA Regulation;

> Teamwork;

> Communication in practical instruction;

> Stress;

> OJTI profile, role, rules and responsibilities;

> Learning problems;

> Recovery of the trainee;

> Motivation; and

> Behavioural theories.

CHAPTER 4

PHO - OPERATIONAL QUALIFICATION PROGRAMME

4.1 Aim and objective of the Programme

The purpose of this program is to establish standardized procedures to be used during the process of operational qualification in Air Traffic Control of professionals coming from training schools or from other locations to meet the specificities of the regional where they will provide their services.

In a previous chapter, the lack of standardization in CTA's training was highlighted. Therefore, the PHO is a program that aims to standardize this training through a method and instruments specially developed to meet the System's needs. It is currently being applied in the System on an experimental basis.

The PHO was prepared by the CINDACTA II Training Team composed of ACC-CW, APP-CT and TWR-CT instructors. They are:

- 1° TEN QOECTA **Paulo** Roberto Soares da Silva

- SO BCT Wagner **Alvarez**;

- SO BCT Mario Luiz Martins **Quintella**;

- SO BCT **Leandro** Pereira da Silva;

- 1S BCT **Joelson** Antônio Rosa Darte;

- 1S BCT Sandro Alves da **Cova** e

- 3S BCT Leandro **Menezes** Rodrigues.

4.2 Responsibilities

In the elaborated documentation, the responsibilities of each sector are pointed out, as well as those attributed to the instructors, based on what the PHO documents recommend.

For the instructors it is incumbent:

- Comply with the schedule designated by the Instruction Subsection for theoretical and practical instruction in accordance with the established syllabus;

- To appraise the trainee's performance, using the respective Assessment Form and filing it in a specific folder;

- Brief the trainee(s) before taking up duty at any of the ATC unit's operational positions;

- Carry out debriefing at the end of each instruction given, correcting certain actions in the performance of the Trainees that may compromise their evolution, as well as elucidating initiatives and procedures that have been carried out during the internship;

- Inform any adverse situation or behaviour involving the Trainee directly to the respective Tutor/Instructional Subsection, so that steps may be taken to ensure that the traineeship is not jeopardised; and

- Fill out the instruction/evaluation form correctly, as well as the required signatures, according to the evaluation manual.

To fulfil the responsibilities listed above, it is necessary not only to know the PHO documentation in detail, but also to obtain specific instructional knowledge and skills that today, most instructors working in the field do not have.

Air traffic controller practical training instructors must have pedagogical knowledge in order to be able to properly plan a practical lesson, to have interpersonal skills to adequately interact with the student besides being able to properly evaluate their trainees among other actions related to instruction. Finally, this knowledge must be associated to the application of the procedures provided in the OHP.

3.4 Standardisation

In accordance with ICA 100-18, Operational Traineeships shall be delivered by applying the Standardised Programme of Instruction approved by the Regional Commander responsible for issuing the CHT.

The Operational Traineeships shall be applied and supervised by the ATC body. The Technical Qualification Certificate shall be granted to the controller who:

a) Hold an air traffic control licence;

b) Have a valid aeronautical medical certificate or health card;

c) Successfully conclude the theoretical phase of the specific operational internship for qualification in the Organ in which he/she will exercise the operational function, demonstrating knowledge of the items listed in ICA 100-18 for the intended CHT (technical qualification card);

d)   Successfully complete the practical phase of the specific Operational Internship for qualification in the body in which it will exercise its operational function; and

e)   It is approved by the Operational Council.

Therefore, every transferred Air Traffic Controller who will perform his functions in an Air Traffic Control Unit, regardless of qualification, must comply with the same program that involves a visit to the administrative/operational facilities related to the Unit, such as INFRAERO, Air Base, GCC (Communication and Control Group), PAR and to the Airport Site, visiting runways, navigation aids and others.

A theoretical phase must be completed, in which the controller must demonstrate knowledge of all the items foreseen in ICA 100-18, specific for qualification in the control body in which he/she will exercise the operational function.

The PHO of each Organ should contain all the items foreseen in ICA 100-18, and the respective publications that deal with the knowledge required for the performance of the function. It is also important to include Notions on Aircraft Performance (civil and military).

In the theoretical phase of the operational internship, contents related to the equipment belonging to the control body that the professional will belong to must be contemplated.

All the equipment that the trainee may use in his professional activity must be approached, with the respective descriptions, not only passing through a theoretical instruction, but also having to handle them under the instructor's supervision.

All this work should be coordinated by the Instruction Subsection, in order for technicians to provide general knowledge about the necessary equipment and navigation aids.

This work foresees a practical phase (simulator and/or real), which is divided into 3 (three) sub-phases where the minimum workload for the operational internship is foreseen in ICA 100-18, depending on the case to which it is destined.

In the simulated practice, the student with the knowledge of all theoretical part pertinent to the Control Body, may apply the knowledge in Low Cost Radar Simulator (SRBC). The workload of this practice may represent up to 30% of the hours planned for the internship.

In the actual practice, the student, with the knowledge of all the theoretical part and simulated practice, if any, pertinent to the control body, must apply it in the operational body. The instructor must keep permanent attention to the internship and check the instructions being issued by the student and correct them whenever necessary. If he/she judges it pertinent, the instructor may interrupt the internship.

After the end of the theoretical phase, before starting the practical phase, the PHO recommends that the trainee should be introduced to the Body to which he will belong as a controller, so that he can make the necessary connections for a better understanding of the link between theory and practice to come. The authors of PHO point out the importance of this environment for an effective work in the cognitive, affective and psychomotor domains of learning to start the work in the initial practical subphase.

The trainee must be instructed to observe the most important points of the traffic dynamics in the respective organ, relating the content of the operational model and the organ's operations manual to the practice under the supervision of an instructor. The trainee must also undergo training in the equipment to be used during the practical internship. The workload provided in the PHO documentation for this training is sixteen hours (two days, eight hours daily).

The practical phase should be divided into three sub-phases:

a)   1ª Practical (Initial) Subphase;

b)   2nd Subphase Practical (Intermediate) and

c)   3rd Subphase Practice (Final).

In the Initial Practice Subphase, the student must be able to recognize the procedures or actions that he/she has been taught and the workload in this subphase corresponds to 15% of the minimum necessary for qualification in the category in which the trainee is training, as provided for in ICA 100-18.

At the end of this sub-phase, the PHO foresees that the student must be submitted to one (01) practical assessment in each operational position in the TWR (control tower) or APP (approach control centre) and two (02) assessments in the ACC (area control centre), conducted by different instructors. Upon receiving a pass in the evaluations, will move on to the Intermediate Subphase.

In the 2nd (Intermediate) Practical Subphase, the workload corresponds to 55% of the minimum required for the category in which the trainee will be training, as provided for in ICA 100-18.

At the end of this sub-phase, the Operational Qualification Program foresees that the trainee shall be submitted to 02 (two) practical assessments in each operational position existing at TWR or APP and 03 (three) assessments at ACC in different positions, carried out by different instructors. When he receives a "pass" in the evaluations, he will pass to the final sub-phase.

In the 3ª Practical Subphase (Final), the workload shall correspond to 30% of the minimum necessary for the trainee to be qualified in the category in which he/she is doing a traineeship, as provided for in ICA 100-18.

At the end, the student must be submitted to 02 (two) practical assessments in each operational position existing in the TWR or APP and 05 (five) assessments in the ACC carried out by different instructors. Upon receiving an "A" grade in the evaluations, he/she will be submitted to the Operational Council.

The programme also provides that, after the assessments in the sub-phases, if the trainee obtains an unfit assessment in at least one assessment in the respective stage, he or she must complete a further 10% of the time allocated for this sub-phase, emphasising the operational position in which he or she obtained a negative assessment, with a view to improving, and must then repeat the assessment. If he/she is assessed as apt, he/she will continue the internship and proceed to the next subphase. If he/she continues with an unsuitable evaluation, he/she must be submitted to the Operational Council at the end of each sub-phase that has not been successful.

PHO reinforces that, in case of force majeure, the assessment carried out by only one instructor may be accepted. However, it is important that this fact is informed in the first deliberation of the Operational Council after the assessment carried out by only one instructor.

After going through the above-mentioned stages, the trainee must go through the actual practical internship, which consists of performing, directly in the operational positions of the ATC (air traffic control) organ following the guidelines contained in the OI Booklet (instruction order), according to the internship schedule created and supervised by the SSIATO (air traffic control instruction subsection) of each regional.

The PHO of each ATC Unit shall contain in this stage all the operational positions of the ATC Unit in question, with the description of the operational profile desired from the controller in the operational position, plus all the items that the trainee shall be instructed and assessed (items from the instruction and assessment sheets).

It is important to remember that, throughout the programme foreseen by the PHO, the Instructor must emphasise the operationalised objectives established in each IO, and should the need arise, the Instructor must review items given in theoretical classes and in an appropriate place, as well as inform the SSIATO of such procedure.

In the general considerations of the PHO there are other elements that should be followed by the Operational Bodies applying the program, which help in the standardization of instruction procedures. They are the following:

- The student shall always be accompanied by an Instructor when in the operational position;

- The internship will consist of two phases (theory and practice), divided into sub-phases (initial, intermediate and final) with the duration foreseen in ICA 100-18, according to each qualification;

- Instructors must use the evaluation criteria and standardization for filling in the Instruction / Evaluation forms defined in the operational evaluation manual;

- The instructor shall substitute the student in more complex situations than those required in the OI (orders of instruction), as well as in situations he/she considers necessary;

- At the beginning of the practical training period in the Operational Unit, the instructor must brief the trainee on the activities that will be developed, in accordance with the OI booklet, and clarify any doubts;

- All IOs must be executed according to the guidelines of the Instruction Subsection, as they were prepared with an increasing level of complexity and with the purpose of approaching all the necessary programmatic content so that the ATCO performs its functions with quality and safety;

- The minimum workload required by ICA 100-18 for each qualification was distributed considering each IO with 1 hour and 30 minutes duration. However, the IO will be considered valid with at least 1 hour, and may be extended up to 2 hours maximum, if a productive operational situation occurs for the internship;

- The Instruction Subsection shall monitor the process in order to account for the internship hours so that the minimum workload foreseen is reached. If the minimum workload is not reached, it must be supplemented with extra IOs from the final subphase;

- At the end of each IBO, the instructor must debrief the student to point out the positive points and the points that need improvement in his/her performance, making comments and recommendations, whenever possible citing the norm that deals with the subjects that need further study to obtain a better performance in the internship.

In its final provisions it also emphasizes that the Operational Proficiency Programme (OPP) shall be applied in conjunction with the Operational Evaluation Manual and the Instructions Booklet of the respective ATC Agency.

4.4 Conclusion on PHO

The PHO represents a valuable tool for standardizing practical instruction for air traffic controllers and was developed taking into consideration the current legislation as well as the educational and psychological theories with a strong technical foundation by the professionals involved.

The military who developed the Programme took the instructors' course (CTP001 and/or CTP006) with the participation and/or coordination of military from the ICEA (Institute for Airspace Control) who later helped in the pedagogic basis and contributed with pertinent information in the construction of the PHO instruments.

When preparing the operational qualification programmes for TRW, APP and ACC, the Training Order Book and the PHO Operational Evaluation Manual, the team considered and used several theories, contents and pedagogical concepts such as Benjamin Bloom's taxonomy of educational objectives (1972), Elizabeth Simpson's psychomotor domain (1972), theories of educational planning and evaluation, learning processes as well as teaching methods and techniques, all permeated by concepts from the area of psychology, as they are directly related to CTA's instruction and professional activity.

After making an analysis of the Program, it is extracted a knowledge base which are necessary to put in practice the actions specified as fundamental for the flight controllers instructor performance. From this Program, it is argued that for the standardization of the instruction in the Operational Bodies, it is necessary previous ability and knowledge of specific nature that allow the instructor the understanding, interpretation and application of the concepts and actions determined in the PHO. Thus, it is believed that the success of the instruction depends on the instructor, he must have a previous pedagogical knowledge base which allows him to act in the teaching-learning process in an efficient and effective way, besides transmitting knowledge and experiences acquired along the years of professional performance.

Reinforcing the argument about the need of a pedagogical knowledge base for the instructor, we recall the survey done about the current practical instructor's profile in chapter 2 of this thesis, in which it is evident that the current profile of this professional is not consistent with the desired needs for the good functioning of the System as a whole.

The new profile should have a base of technical knowledge and skills as well as the ability to teach them, i.e. it is necessary to add pedagogical content in the practical instructor's training.

It is important to note that the PHO was implemented in CINDACTA 2 and is being implemented in CINDACTA 3 and progressively will be in other CINDACTA.

Thus, it is clear that there is an urgent need for a training course for practical instructors of air traffic control that meets the demand of the System and in line with the information and tools necessary for the standardized application of instruction as provided in the PHO.

Thus, the next chapter of this thesis will present the basic content, which should represent the minimum curriculum to compose the air traffic controller practical instructor course.

CHAPTER 5

BASIC PEDAGOGICAL CONTENTS TO BE TAKEN INTO CONSIDERATION WHEN TRAINING THE PRACTICAL INSTRUCTOR

The objective of this chapter is to propose a minimum pedagogical knowledge base so that the ATC practical instructor can accomplish his task which is the operational training of the air traffic controller.

The most relevant pedagogical knowledge he/she must acquire is mainly related to planning, which aims to analyse the teaching demands in order to establish the objectives, contents and skills to be taught, and with that to structure the teaching plan, put it into practice, evaluate it and finally improve it. Besides planning it is necessary that the instructor has knowledge about the functioning of the learning process, including knowledge about learning theories, so he/she can better plan and conduct his/her activities to maximize the results along the instruction process. Additionally, the instructor must receive training on how to develop the evaluation process in the training, in light of the proposed objectives for teaching and learning of the instructed.

This pedagogical knowledge base is described in more detail below.

5.1 Teaching planning

For every human activity that aims at a degree of efficiency or intends to achieve certain objectives, it is necessary to plan. We do not act without first trying to foresee actions to overcome possible obstacles in our paths. In relation to teaching, this is no different.

The definition of planning is brilliantly described by Mattos (1968, p.140), a renowned author in the field of education, when he says that teaching planning is the "intelligent and well-calculated forecasting of all the stages of school work involving teaching and student activities, so as to make teaching safe, economical and efficient."(p.140)

Teaching planning is divided into three phases, preparation, development and improvement. The preparation phase is subdivided into six stages and the evaluation phase into three stages.

According to Turra (2001), there is still an extremely important procedure to be carried out before planning begins, which is the knowledge of reality. A kind of diagnosis of the target audience and instructional conditions to be checked to dimension the possible problems to be encountered during the teaching-learning process and verify the level of cognitive, affective and psychomotor domain in which the students of the course or internship in question find themselves.

5.1.1   Knowledge of reality

Knowledge of reality is the verification of the target audience, its workplace and the environment available for the class. In this stage it is done a survey and from this it is done the diagnosis to prepare the instruction. This way, the instructor has a reference to elaborate a teaching plan, supported by real and significant causes.

5.1.2   Preparation

It is at this point that the objectives that students should achieve at the end of the instruction are established. Then, this phase sets out to verify which contents are necessary for students to achieve the proposed objectives, selects the best teaching techniques (teaching procedures) to be used to work on the contents, determines the resources necessary to operationalise the lesson and, finally, chooses the best way to assess the student (assessment procedures) and verify whether they have achieved the proposed objectives.

5.1.3   Development phase

The teaching plan is put into action, it is the operationalization of the plan. The roles of the instructor and the student in the implementation of the plan are highlighted.

5.1.4   Fine-tuning phase

This assessment adopts a different meaning from the assessment of learning, as it has a broader significance with a view to replanning. It works in a way that feeds back into the planning system.

Figure 1: Phases of teaching planning (Turra, 1975, p. 26)

5.2 Taxionomy of educational objectives

The teaching objectives are descriptions of planned results for a teaching-learning situation which will be observable in the learner.

The formulation of teaching objectives is a task of great importance because it is the objectives which, in addition to providing the teacher with the guidelines for preparing his class, allow him to predict the results he wishes to achieve at the end of the instruction. In other words, teaching objectives indicate the behavioural changes to be achieved by the students.

5.2.1 Location of objectives in the teaching planning process

The development of teaching objectives is the most important step in the planning process. It is done at the very beginning, before thinking about content, techniques, resources and assessment. Thus, the first step in any planning is the formulation of objectives. It is around them that the whole planning process revolves, because it is only after their elaboration that the instructor must plan the subsequent steps.

5.2.2 Ways of classifying educational objectives

There are a variety of classifications of objectives, but the one most suited to practical instruction, is the classification under two approaches (Turra, 1975):

> As to the degree of coverage; and

> Regarding the domains and levels of learning.

SCOPE	GENERAL
	SPECIFIC
	OPERATIONAL

Source: Turra, 1975

DOMAINS AND LEVELS OF LEARNING	COGNITIVE
	AFFETIVO
	PSICOMOTOR

Source Turra, 1975

It is important to remember that one classification does not exclude the other, that is, the same objective always belongs to a certain domain of learning, while having a level of comprehensiveness.

With regard to the degree of coverage (Turra, 1975):

- General objectives are more complex, their learning outcomes are achievable over longer periods of time, they relate to the final behaviours that the student is expected to display at the end of the course or internship.

- Specific objectives are simpler, more concrete, achievable in less time and make observable performances explicit.

- Operational objectives are those which describe a behaviour to be shown by the student at the end of the lesson or training day, also presenting the conditions under which this behaviour should manifest itself and the criteria or standard by which it can be measured.

As for the learning domain (Turra, 1975):

- Cognitive objectives are those related to knowledge, concepts, ideas, principles and mental abilities.

- Affective goals are those that refer to attitudes, interests, values and appreciations.

- Psychomotor objectives are those aimed at developing manipulative or motor skills.

5.2.3 Objective classification system

Every teaching objective brings a learning level which can be defined as the degree of complexity or depth with which a subject will be covered.

Of the existing classification systems, the best known and most widely used is the "Taxionomy of educational objectives", organised by Benjamin Bloom and team.

The taxionomy is divided into (BLOOM et al,1972):

> cognitive domain;

> affective domain; and

> psychomotor domain.

Each domain is subdivided into levels.

Benjamin Bloom (1972) only completed his studies of goals in the cognitive and affective domains, for the psychomotor domain Elizabeth Simpson's (1971) classification will be presented.

> Cognitive domain

For each domain, Bloom established an integrating principle or organizational criterion. For the cognitive domain, the criterion chosen was that of increasing complexity, according to which simpler learning outcomes are achieved until more complex levels are reached. According to this principle, a level encompasses those which precede it.

For the cognitive domain, the taxonomy presents six levels (BLOOM et al,1972):

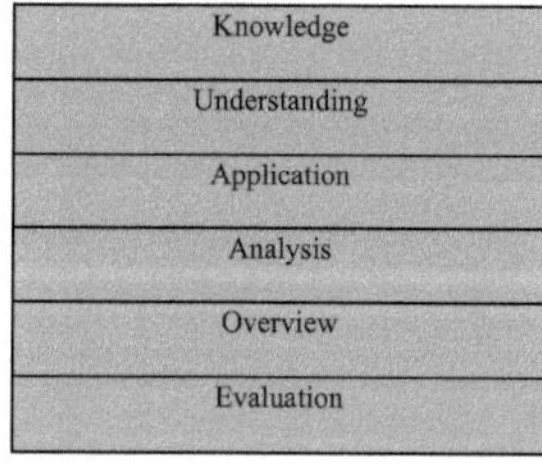

<u>Knowledge (Cn)</u>

It is the most elementary level of this domain and can be defined as the recall of previously learned material. This may involve recalling a long range of material, ranging from specific facts to complete theories, but all that is required is the recall of the appropriate information. At this level no conclusions or applications are required of the learner.

At this level, the student:

- defines terms, citing their attributes, properties or relationships;

- evokes specific facts;

- identifies symbols, rules;

- describes processes, directions and trends of phenomena in relation to time;

- knows classes, categories, judgement criteria, methodologies, principles, theories and structures.

Understanding (Cp)

At this level, the student, in addition to understanding the exact meaning of the subject, understands the relationship between its components and how and why its effects are produced.

In understanding, the student:

- reorganises a material, giving it a new disposition and relating it to their experiences (interpretation);

- draws conclusions, makes inferences and predicts consequences (extrapolation).

Application (Ap)

It refers to the ability to use learned material in new, concrete situations without being suggested which abstractions to use or being taught how to use them in that particular situation.

At this level, the student:

- applies rules, methods, concepts, principles, laws and theories.

Analysis (An)

Analysis refers to the ability to break down a material into its component parts and relate them in order to perceive its organisational structure.

At this level, the student:

- identifies all the constituent parts of a material;

- establishes relationships between the elements found;

- identifies the organisational structure used.

Synthesis (Si)

Synthesis is the bringing together of elements and parts to form a whole. It is a process of working with elements and parts in order to combine them to form a new structure.

This level provides the student with greater opportunities to develop creative behaviour, although they need to work within the limits imposed by certain problems, materials or theoretical and methodological frameworks.

At this level, the student:

- transmits ideas, feelings and experiences to other people;

- produces plans;

- makes discoveries and generalisations.

<u>Assessment (Av)</u>

The evaluation level is defined as the process of judging the value of ideas, works, solutions, methods, materials, carried out for a specific purpose.

Evaluation can be carried out using internal or external criteria.

At this level, the student:

- makes judgements based on internal criteria such as logical accuracy, clarity, consistency, etc.

- makes judgements based on selected external criteria.

> Affective domain

The affective domain, although of great importance, is relegated to a secondary plan, where the main concern is to develop the cognitive domain. Bloom established five levels for the affective domain. The integrating criterion selected was that of "internalization", which can be understood as "the individual's acceptance of attitudes, codes, principles or sanctions, which become a part of himself, in the formation of value judgment or in the determination of his conduct". Bloom (1972).

This behaviour is partly due to:

- difficulty in formulating affective goals in behavioural terms;

- difficulty in assessing them;

- slow achievement of them.

The levels presented are as follows:

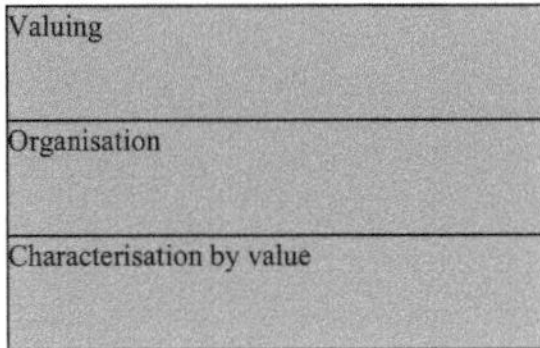

Reception (Ac)

Willingness, on the part of the student, to pay attention to particular phenomena or stimuli.

Answer (Re)

At this level, the learner not only pays attention to a phenomenon but also reacts to it in some way.

Valuation (Va)

It refers to the value that the student/trainee attributes to a given object, action, phenomenon or behaviour.

Organisation (Og)

It refers to the bringing together of different values, comparison between them and the beginning of the

elaboration of an internally consistent value system.

Characterisation by a value or complex of values (Cv)

At this level, the learner has a value system, controlling their behaviour.

> The psychomotor domain

Elizabeth Simpson established five levels for the psychomotor domain:

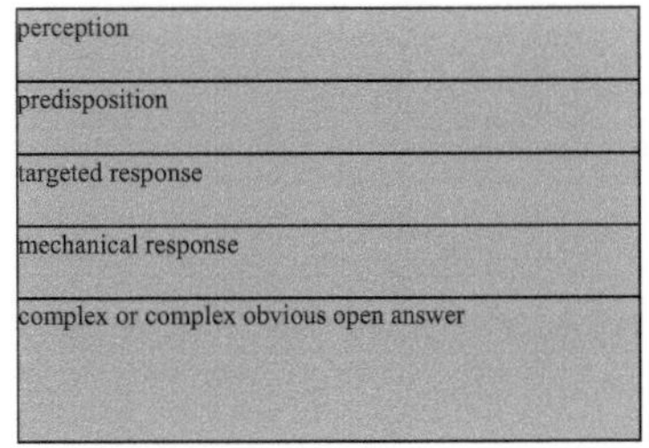

Perception (Pe)

It refers to the awareness of the situation, through the senses.

Predisposition (Pr)

It refers to the mental, physical and emotional adjustment to action.

Targeted response (Ro)

Beginning of psychomotor action. Consists of the action of the individual under the guidance of an instructor.

<u>Mechanical response (Rm)</u>

At this level, the individual has already developed some degree of skill in performing the action.

<u>Complex open response or obvious complex response (Rc)</u>

It consists of the performance of a motor act with a high degree of skill.

It should not be forgotten that one of the major advantages of objectives is that they assist in assessment. If the objectives describe what we want the students/trainees to learn, we will only know if they have actually learned what we are trying to teach if we assess them properly.

Thus, the evaluation should verify the behaviour in the scope, level and domain specified in the objectives.

5.3 Learning

Practical-operational instruction assumes considerable importance, since it is the link between the training given at the Air Force Academy and the effective assumption of duties. The knowledge about the learning process allows the Instructor to plan and conduct his activities in order to maximize the results of the process.

5.3.1 Concept of Learning

Learning can be briefly defined as the way in which human beings acquire new knowledge, develop skills and change behaviour. This process occurs through interaction with the environment, with other individuals and with experience.

5.3.2 Practical learning theory

The way in which man learns and the ways in which one can teach have been the subject of study since ancient times. It is important to note that most of the theories

address the learning that accompanies child development. Only recently has the

adult education has come under the spotlight.

In the case of SISCEAB instruction, for its peculiarities, the use of Piagetian and social interactionist approaches are more appropriate, despite the technical character of the System activities. Such option is due to the belief that the instructor needs to have subsidies to plan his activities, taking into account the phenomena between the instruction and the effective condition of use, by the trainee, of the knowledge in a work situation.

The Piagetian approach is called constructivism, because for Piaget knowledge is not received, it is necessary that the individual builds it internally. In this theory, man's adaptation to the environment occurs through processes he calls assimilation and accommodation. Assimilation consists in the incorporation of environmental elements by the individual. The accommodation is the process of modification that operates in the structures of thought that the subject already had, due to the assimilation of new knowledge. In this context it is important to differentiate experience and knowledge, because the latter presupposes the internal organization of experience.

While Piaget focused on the biological aspects of development, in the social interactionist approach the focus is on social relationships that promote the development of the individual. From this point of view, the relationship between man and his environment is not natural, nor is it a simple adaptation, involving very complex interactions. In the relationship with the environment around him/her, the individual uses socially constituted instruments and signs.

Vygotsky, father of the social interactionist theory, considers that every psychological function is developed first at the level of relations between individuals and, subsequently, in the individual himself. Thus, the child knows the world through relationships with other people, which allow him/her to internalise reality.

In the social interactionist perspective, learning arouses and drives the development and one should work with the indicators of proximal development, which reveal the ways of acting and thinking which are still being developed and which require guidance from the other in order to be realised. Vygotsky distinguishes three zones of development:

- Potential in which the subject meets the necessary conditions to perform a given task

- Proximal the subject is able to perform the task but with the assistance of a mediator, that is, with help

- The learner is able to carry out the task without any help).

In the constructivist theory, according to Piaget, every time a person comes into contact and interacts with new knowledge, whatever it is, practical or not, he or she goes through the process of assimilation, often similar to a mental confusion, until this knowledge is progressively "accommodated" - accommodation - which would be where the contents or experiences would make some sense to him or her because they are related to things he or she already knows. Going through this process, the subject reaches a new stage of equilibrium.

Figure 2 illustrates the mental processes that the trainee goes through during a learning/ instructional moment according to the two theories presented above.

In figure 2 the concepts of the theories were used in relation to the mental processes of learning through which the trainee passes for better visualization of the proposal. However, it is essential to emphasize that at each new learning, the subject will go through this path.

It is, therefore, a continuous process of development, since the comfort attained by achieving dexterity in a given procedure will be broken at the moment when he/she realises that he/she does not have the subsidies to solve a new situation.

DEVELOPMENT PROCESS
IN PRACTICAL-OPERATIONAL INSTRUCTION

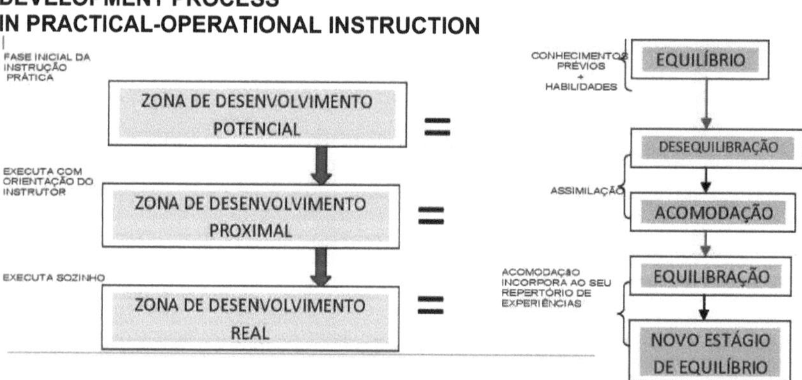

Figure 2: Associated theories for practical instruction

5.4 Evaluation

For many teachers, assessment is, still today, a simple activity of "giving grades to students", or establishing "questions" to embarrass them. For students, assessment generally represents a "series of obstacles to be overcome", as the only condition for approval in a course.

In fact, assessment is the thermometer that allows confirming the state of the elements involved in the teaching-learning context, i.e. student, instructor, teaching methods and techniques, learning assessment process and instruments. It plays a highly significant role in instruction and it can even be said that assessment is the soul of the instructional process.

The different functions it assumes, play a decisive role and none of them exclude the evaluator and the evaluated from the commitment of being their own decision agent and responsible for the instruction process.

Evaluation is a complex process and not a simple awarding of marks. It begins with the formulation of objectives and requires the development of means to obtain evidence of results. The interpretation of these results shows to what extent the objectives have been achieved and allows a value judgement to be formed.

5.4.1 Modalities and functions of Evaluation

According to Bloom (1971), assessment is classified into three functions/modalities:

Chart 11 - Modalities and Functions of Evaluation

MODALITY	FUNCTION	
DIAGNOSTICS	DIAGNOSTICS	It aims to verify the presence or absence of knowledge and skills that are prerequisites for new learning experiences.
TRAINING	CONTROL	Inform the instructor and the student about the learning outcome. Verifies the achievement of the objectives proposed for the instruction. If they have not been achieved, it allows the instructor to make the necessary corrections through pedagogical interventions throughout the teaching process.
SOMATIVA	CLASSIFICATION	Its function is to classify the students at the end of the unit or course, according to the levels of achievement presented. The classification is carried out according to the performance achieved, using the planned objectives as a parameter.

The goal that every instructor sets for themselves is for students to score well on tests. However, only this analysis is not a guarantee that learning will be effective. It is necessary to verify how much the students/trainees already know about a certain subject or procedure that will be approached in order to adequate the content and the activity to be worked with the group of trainees. It is important to remember that there must also be continuous verification of the achievement of the objectives proposed by the instruction so that the student does not have deficiencies and difficulties at the end of the internship.

Finally, an evaluation must be carried out in order to classify the student, but we must still be concerned with how to elaborate the appropriate measuring instrument to verify the achievement of the intended objectives.

CHAPTER 6

6 PROPOSAL FOR A MINIMUM CURRICULUM FOR THE SISCEAB AIR TRAFFIC CONTROLLER OJTI

6.1 Minimum Curriculum Proposal

Based on the analysis of the training courses abroad, relevant aspects to be considered in the course proposal were identified.

It must also be taken into account the set of technical knowledge that the PHO requires, since the instructor must know and master them in order to be able to apply the Operational Qualification Programme as it is foreseen.

Finally, pedagogical contents should be included in the minimum curriculum proposal in order to make possible the operationalization of the instructions as they are foreseen, since pedagogical knowledge is necessary to interpret and apply the instruction activities. Moreover, these contents will help the instructor to improve his ability to transmit the technical knowledge, giving him specific skills for practical instruction.

6.2 Minimum Curriculum Vitae

6.2.1 Structural design of the course

The On-The-Job Training Instructor Course for Air Traffic Control - ATM 050, aims to provide students with theoretical and practical learning experiences that enable them to exercise the function of practical-operational air traffic control instructor within the Brazilian Airspace Control System.

The course is structured through three technical areas proposing to be an interdisciplinary course involving air traffic control, pedagogy and psychology.

In this course, specialised air traffic control instruction subjects will be covered for practical instruction.

This Minimum Curriculum establishes the contents of the theoretical phase of the course and it is the responsibility of the ICEA together with the various operational bodies to develop a Teaching Unit Plan that defines the practical phase of the OJTI internship meeting the specificities of each regional or operational body. This document must be in line with the theoretical phase of the course and may in no way be dissociated from the content provided in the Minimum Curriculum and Teaching Unit Plan of the theoretical phase.

6.2.2 Specific performance standards

a) Map the competencies of the air traffic controller;

b) Plan theoretical and practical instruction for the air traffic controller;

c) Evaluate air traffic control trainee students;

d) To provide instruction for air traffic controllers in theory classes and in simulated or OJT training;

e) To make appropriate use of assessment tools for student trainees;

f) Prepare assessment instruments appropriate to the instruction to be delivered;

g) Use knowledge of the psychology of learning in practical instruction; and

h) To build the technical capacity of air traffic controllers.

6.2.3 Student profile

Students of the On-The-Job Training Instructor Course for Air Traffic Control - ATM 050 must present the following characteristics:

a) Be a military graduate of the air traffic control specialty or similar civilian;

b) Have been operating in air traffic control for at least 2 years;

c) have been operating in the position in which it is intended to instruct for at least 6 months; and

d) To have been selected by the operational body in which you work or INFRAERO.

6.2.2 Aim, general objective and duration of the course

- Purpose of the course

The On-The-Job Training for Air Traffic Control course aims to train practical-operational instructors that meet the specific needs of air traffic control in SISCEAB.

- General objective

To enable air traffic controllers to provide quality practical instruction in an operational control body; and

To train the air traffic controller OJTI (On-The- Job training instructor) in line with international standards.

- Duration of the course:

It is suggested that the course has a duration of 45 (forty-five) days, 30 days in the distance learning modality and 15 days or 120h/class in the face-to-face modality. It will have to be taken into consideration in the assembly of the course face-to-face phase, the times related to the following activities:

a) Administrative briefing;
b) Evaluation and
c) Programming Flexibility.

It is important to emphasize the importance of organizing a supervised internship that complements the course. Such internship for practical instruction should have theoretical support and briefing in the operational body to which the instructor intends to minister instruction.

6.3   Curriculum Content

6.3.1   Course overview

Disciplines:

- Operational safety

- Operational qualification

- Technical areas integrated into the operational body

- Skills and training

- Teaching planning

- Evaluation

- Operational qualification programme

- Psychology and learning

Specific objectives to be achieved in the disciplines:

❖ Operational safety

   a)   Assess the philosophical principles governing operational safety (Av);

   b)   Apply the concept of Operational Safety Management System (Ap);

   c)   Identify flight safety applied to air traffic control (AN);

d)   Relate the civil responsibility in the performance of the practical operational instructor well

       as in the performance of the air traffic controller and control activity trainee (Av);

e)   Relate specific air traffic control standards to the preservation of

       operational safety (Av).

EMPHASISE:

1)   Philosophical principles of operational safety;

2)   Operational Safety Management System (OSMS)

3)   Flight safety applied to air traffic control; and

4)   Civil liability;

5)   Specific air traffic control standards.

❖   Operational qualification

a)   Describe the process of operational clearance (Av);

b)   Discriminate the skills required for the performance of the air traffic controller

     (An);

c)   Relate operational clearance to flight safety (Av).

EMPHASISE:

1)   operational qualification process

2)   skills necessary for the performance of the air traffic controller

3)   operational clearance and flight safety.

❖   Technical areas integrated into the operational body:

a) Describe the importance of the integration of technical areas for the proper functioning of air traffic control (Av);

b) Justify the need for integration of different specialties for the preservation of operational safety (Av);

c) Explain the operational training of the different integrated technical areas (Cp).

EMPHASISE:

1) Integration of technical areas for the proper functioning of traffic control air;

2) Integration of the different specialities to preserve operational safety;

3) The operational training of the different integrated technical areas of the control.

❖ Skills and training

a) Relate competence and training (Av);

b) Justify the importance of the training (Av);

c) To map the competencies required for the performance of the air traffic controller (Si);

d) To identify the management by competencies as a way of enhancing the performance of the professional activity of the air traffic controller (Av);

e) Relate multiple intelligences to the development of skills for the performance of competencies (Av).

EMPHASISE:

1) Skills and training - Introduction;

2) Staff training and capacity building;

3) Mapping of competencies required for performance in air traffic control;

4) Management by competencies;

5) Multiple intelligences

❖ Teaching planning

a) Elaborate teaching plans (Si);

b) Explain domains and levels of learning (Av);

c)	Writing teaching objectives (Si);

d)	Select contents for instruction (Si);

e)	Select the best teaching technique for air traffic control (Av) instructions;

f)	Select the resources appropriate to the teaching techniques (Si);

g)	Select the best assessment procedures for the type of instruction chosen (Si);

h)	Apply the teaching planning (Ap);

i)	Evaluate the teaching plan (Av); and

j)	Describe the TRAINAIR programme and its methodology of course development. EMENT:

1)	Teaching planning;

2)	Domains and levels of learning;

3)	Drafting of teaching objectives;

4)	Content selection;

5)	Teaching techniques for practical and theoretical instruction;

6)	Resources;

7)	Evaluation procedures;

8)	Execution of the teaching plan;

9)	Evaluation of the teaching plan; and

10)	TRAINAIR programme and methodology.

❖	Evaluation

a)	Describe diagnostic, formative and summative assessment for theoretical and practical air traffic control instruction(Av);

b)	To formulate theoretical and practical Evaluation instruments for instruction aimed at air traffic control (Si);

c)	Use the assessment instruments foreseen in the PHO (Ap);

EMPHASISE:

1)	Diagnostic, formative and summative assessment for theoretical and practical air traffic control instruction;

2) Theoretical and practical assessment tools; and

Assessment instruments provided for in the PHO;

❖   Operational qualification programme

a)   Identify the theoretical and practical simulated and real phases of PHO (Ap);

b)   Apply the assessment foreseen in the PHO (Ap);

c)   Use the assessment instruments foreseen in the PHO (Ap);

d)   Handle the control body equipment foreseen in the PHO (Ap);

e)   Apply supervised internship as provided in the PHO (Ap).

EMPHASISE:

1)   Theoretical phase of the OWP and simulated and real practice of the OWP

2)   PHO simulated practical phase

3)   Actual practical phase of the PHO

4)   assessment provided for in the PHO

5)   Assessment instruments provided for in the PHO

6)   Equipment of the control body foreseen in the PHO

7)   Supervised internship provided for in the PHO

❖ Psychology and learning

a)   Explain the cognitivist, socio-interacting and behaviourist (An) approaches;

b)   To relate educational approaches to air traffic controller (Av) training;

c)   Explain the factors intervening in the practical operational learning of the air traffic controller (Cp); and

d)   Analyse the trainee's learning process (An)

e)   To identify the psychological factors present in the teaching-learning process that can both facilitate and

compromise the instructional activity (Cp);

f)   Differentiate the possible forms of instructor performance facing the cognitive and emotional issues during

the practical instruction of the air traffic controller (Cp); and

g)   Explain the influence of interpersonal relationships on theoretical and practical air traffic controller (Av) instruction.

h) Explain the impacts of stress on the performance of air traffic control activity and practical instruction.

EMPHASISE:

1) Cognitivist, socio-interacting and behaviourist approaches;

2) Educational approaches and the training of air traffic controllers;

3) Factors intervening in the practical operational learning of the controller;

4) Ways of acting of the instructor facing cognitive and emotional issues during instruction;

5) Interpersonal relationships in theoretical and practical controller instruction;

6) Trainee learning process; and

7) Stress.

6.4   Evaluation procedures

The assessment procedures for the course shall be as follows:

❖ Evaluation of the theoretical distance phase - An objective online test;

❖ Evaluation of the theoretical classroom phase - a diagnostic evaluation before beginning the classroom phase, performed by an objective test and three theoretical evaluations throughout and at the end of the phase by means of two objective tests and a discursive test.

❖ The student will be considered apt for the practical phase if they obtain a 70% pass in the theoretical phase. The only test that will not be given any value will be the diagnostic test, which will serve only to check the level of knowledge for the beginning of the on-site phase.

6.5   General provisions

❖ For the Complementation of Instruction activities, it is suggested to organize lectures and visits that promote a greater knowledge of the work environment related to air traffic control.

❖ At the end of the course, the student must carry out some important activities for their professional performance which are:

a) Guided tour explaining the various sectors that work in integration with air traffic control;

b) Lecture with information on how practical instructions are given for the training of professionals from each sector that makes up the body;

c) Visit to the working environment facilities including airport and airstrip if applicable.

CHAPTER 7

SUGGESTIONS FOR WORK AND CONCLUSION

7.1 Research and recommendations

The problem studied in this work was the air traffic controller practical instructor training. Through documentary analysis, bibliographical research, questionnaire application and non-structured interview, it was concluded that the training given to this instructor in the System today, although it has been improving the quality of instruction, is still not enough to meet the specificities of the OJTI activity for CTA.

Furthermore, only a small proportion of practical instructors and future trainee instructors have access to CTP006 due to the fact that it is difficult to spare network professionals to undertake the course.

During the analysis of the RICEA, it was found a significant difficulty to identify the information needed for the research because the document does not have any synthetic and objective part that could help in the search for a theme. Thus, it was necessary to read almost the entire document in order to identify a record that was associated with the desired information.

This issue may represent difficulties even for the processing of information by the competent body, as well as for the rapid search for important information to solve problems presented by the network (SISCEAB).

Therefore, the recommendations made by the researcher from this research are:

- The creation of a specific course for practical instructors of air traffic controllers and a certification process for this professional;

- To disseminate the course to the practical instructors as well as knowledge that will assist the instructors in their training; and

The recommendations for the solution presented for the above problems are:

- The organization of a WG composed of a multidisciplinary team with pedagogues, psychologists and air traffic controllers to analyze, alter and approve the minimum curriculum described in the previous chapter, as well as the elaboration of a Teaching Unit Plan and didactic material for the application of the ATC practical instructor course; and

- The development of a space for knowledge exchange and application of courses or phase of courses for OJTIs on a distance learning platform.

7.2 Conclusion

The practical-operational instructor training in air traffic control is an extremely important factor, not only for the adequate training of professionals of the sector, but also as a way to preserve the operational safety during instruction.

In this work it becomes clear the alignment between the regulatory bodies of the practical-operational instruction activity in relation to OJTI training in EUROCONTROL, England, Singapore and the United States (FAA). The several coinciding points converge towards a standardization of instruction and, consequently, towards a standardization of the air traffic controller teaching-learning process, representing an increase in operational safety.

Therefore, the researcher recommends that, following the example of what occurs in the researched countries, it is important that standardization of instruction for the OJTI of controllers be applied in line with what is provided by the main international air traffic control bodies.

It is worth noting that air traffic control activity is of great relevance and responsibility since it deals with human lives.

A controller with deficient training represents risks to operational safety. Therefore, one way to minimize risks is by offering quality training through instructors who are well prepared to perform their function, and who have all the necessary "tools" for their pedagogical performance.

The product of this work presents an outline for the solution of this problem, which is to provide the instructors with the knowledge and skills necessary for the performance of the operational practical instruction activity for air traffic controllers. In chapter 6, there is a draft of a minimum curriculum, basis for the construction of the documentation required by the Air Force for the creation of a course, to be discussed and changed by a working group, for the creation of a specific course for practical instructors of air traffic controllers.

Good air traffic controller training translates into minimising risks in aviation.

REFERENCES

BLOOM, Benjamin S.; HASTINGS, J. Thomas; MADAUS, George F. **Manual de avaliação formativa e somativa do aprendizagem escolar**. São Paulo: Pioneira, 1971.

BLOOM, Benjamin. **Taxionomia dos objetivos educacionais**. Porto Alegre, Ed Globo, 1972.

BOOG, Gustavo G. **Manual de Treinamento e Desenvolvimento**. São Paulo: Makron Books, 1999.

BOMFIN, David F. **Pedagogia no Treinamento**: Correntes Pedagógicas no Ambiente de Aprendizagem nas Organizações. Editora Qualitymark - 2ª Ed. - RJ-2007.

BORDENAVE, Juan Diaz and PEREIRA, Adair Martins. **Estratégias de Ensino- Aprendizagem**. Petrópolis: Vozes, 1997.

BRAZIL. Air Force Command. Department of Aeronautics Education. **IMA 37-10.** Basic Concepts in Teaching: Rio de Janeiro, 1983.

BRAZIL. Aeronautics Command. **IMA 37-8.** Objetivos de Ensino e níveis a atingir na aprendizagem. Rio de Janeiro, 1988.

BRAZIL. Air Force Command. Airspace Control Department **Ordinance 118/DGCEA**, August 11, 2010.

BRAZIL. Aeronautics Command. **ICA 63-11/2010** Structure and Attributions of the Security Subsystem of the Brazilian Airspace Control System.

BRAZIL. Aeronautics Command. **ICA 63-7/2010 ICA:** Attributions of the SISCEAB Organs after the Occurrence of Aeronautical Accidents or Serious Aeronautical Incidents

BRAZIL. Aeronautical Command. **ICA 100-5/2003:** Air Traffic Incidents Investigation

BRAZIL. Aeronautical Command. **ICA 100-18:** Licenses and Certificates of Technical Qualification for Air Traffic Controllers

BRAZIL. Aeronautics Command. **ICA 100-30:** ATC Personnel Planning, 17JAN08.

BRAZIL. Aeronautics Command. **ICA 100-12:** Rules of the Air and Air Traffic Services. 01 April 2009.

BRAZIL. Aeronautics Command. **NPA 142/D0/260608**: Structure and Operation of Instruction Sections of ATS Organs.

BRAZIL. Air Force Command. Operational Qualification Program, 2010.

BRAZIL. Aeronautics Command. Operational Evaluation Manual of the PHO, 2010.

BRAZIL. Aeronautics Command. PHO ACC Operational Qualification Program, 2010.

BRAZIL. Aeronautics Command. Operational Habilitation Program of PHO APP, 2010.

BRAZIL. Aeronautics Command. PHO TRW Operational Qualification Program, 2010.

MATTOS, L. A. **Sumário de didática Geral**. Rio de Janeiro: Aurora, 1968.

BRAZIL. Aeronautics Command. Airspace Control Institute. Aspectos Psicológicos. Apostille of the ATC organs supervisor course; 2001.

CAPELLE CONSULTING. Available at: http://www.capelleconsulting.com/sg. Accessed June 2010.

CIVIL AVIATION AUTHORITY OF SINGAPORE. Manual of Standards: Licensing of Air Traffic Control Personnel. RC: CAAS, 2010. 45 p.

CHIAVENATTO, Idalberto. **Recursos Humanos**. São Paulo: Atlas 1985.

COVEY, Stephen R. **Leadership based on principles**. Rio de Janeiro: Elsevier, 2002.

DALGALARRONDO, P. **Psicopatologia e Semiologia dos Transtornos**. Porto Alegre: Artes Médicas, 2000.

DAVIDOF, Linda. **Introduction to Psychology. São Paulo**: Makron Books do Brasil, 1983.

UNITED STATES. Astac at http://www.astac.co.uk . Accessed June 1, 2011.

UNITED STATES. Federal Aviation Administration. Air Traffic Technical Training. [S.L.], 2005.393 p..

UNITED STATES. Federal Aviation Administration. Air Traffic Technical Training 3120.4L . Available at: <http://www.faa.gov/training_ testing/training/fits/research/media/TAA%20Final%20Report.pdf>. Accessed: 12 Oct. 2011.

UNITED STATES. Cwmbran College. http://www.cwmbrancollege.com. Accessed June 2011.

EUROCONTROL, **ESARR 5: 2002** - Eurocontrol Safety Regulatory Requirement. [S.L.], 2002. 24p.

EUROCONTROL, Air Traffic Controller Training at Operational Units. [S.L.], 2002.

EUROCONTROL. European Organisation for the Safety of Air Navigation. [S.L.], 1999, 274 p.

EUROCONTROL, ATCO Development Training: OJTI Course. [S.L.] 2004, 32 p.

GAGNÉ, Robert M. **Como se realiza aprendizagem**. Rio de Janeiro: Livros Técnicos e Científicos, 1974.

HAMBLIN, A.C. **Avaliação e Controle do Treinamento**. São Paulo: McGraw-Hill, 1978.

HAMMER, Michel James. **Reengenharia**: revolucionando a empresa em função dos clientes, da concorrência e das grandes mudanças da gestão. Rio de Janeiro: Campos, 1994.

INTERNATIONAL CIVIL AVIATION ORGANIZATION. **DOC 9868** : PANS Instruction, [S.L.] 2006.

INTERNATIONAL CIVIL AVIATION ORGANIZATION. **DOC 7192-AN/857**: Training Manual, 2004.

INTERNATIONAL CIVIL AVIATION ORGANIZATION. **RLA/06/901.** Competencies for instruction, 2010.

INTERNATIONAL CIVIL AVIATION ORGANIZATION. **RCC/4 NI/03.** [S.L.], 2010, 90 p.

INTERNATIONAL CIVIL AVIATION ORGANIZATION . **4 NI/03** Communication and Coordinations [S.L.], 2010.

INTERNATIONAL CIVIL AVIATION ORGANISATION . **DOC 9868** . PANS Instruction. [S.L.] 2010.

INTERNATIONAL CIVIL AVIATION ORGANIZATION. **DOC 7192-AN/857**: Training Manual. [S.L.] 2010.

INTERNATIONAL CIVIL AVIATION ORGANIZATION. **RLA/06/901- RCC/4 NI/03** - competencies for instruction. [S.L.] 2010.

LAVILLE, Christian; DIONNE, Jean. **The Construction of Knowledge**: Handbook and Methodology of Research in the Human Sciences .: ARTMED, 2010.

LIBÂNEO, José Carlos. **Didática**. São Paulo: Cortez, 1992.

LINDEMAN, Richard H. **Medidas Educacionais**. Porto Alegre: Globo S/A, 1978.

LURIA, A. R. - **General Psychology Course**. São Paulo: Civilização Brasileira, 1991. Vol. 2 and 3.

MACIAN, Leda Massari. **Training and Development of Human Resources**. São Paulo: EPU, 1987.

MOSCOVIS, Fela. **Interpersonal development:** group training. Rio de Janeiro: José Olympio, 1998.

NÉRICI, Imídeo Giuseppe. **Didática geral dinâmica**. 10. ed. São Paulo: Atlas, 1989.

PENNA, Antonio Gomes. **Perception and Reality**: introduction to the study of perceptive activity. Rio de Janeiro: Imago, 1993.

SANTOS, Thelma Rodrigues. **Oratory**: how to speak in public. 2006. Apostille.

SINGAPORE. Singapore Aviation Academy at http://www.saa.com.sg . Accessed 02 June 2011.

SPECTOR, Paul. - **Psychology in organizations**. São Paulo: Saraiva, 2004.

TOLEDO, Flávio de; MILIONI, B. **Dicionário de Recursos Humanos**. São Paulo: Atlas 1986.

TURRA, Clodia Maria Godoy et alii. **Planejamento de ensino e avaliação**. 11ª ed. Porto Alegre: PUC/EMMA, 2001.

APPENDIX 1: QUESTIONNAIRE APPLIED TO AIR TRAFFIC CONTROLLERS
Questionnaire for air traffic controllers

( ) CINDACTA 1   ( ) CINDACTA 2(   ) CINDACTA 3( ) CINDACTA 4

( ) ICEA(   ) SRPV-SP   ( ) PAME

In the operational body in which you work or have worked:

1-   Historically, how are instructors chosen to deliver the operational internship?

( ) by seniority, the oldest;

( ) the most modern;

( ) the military with more experience in the area of performance

( ) someone who has specific training for operational instruction

2-   Is any specific training compulsory for someone to become an instructor and start teaching in the operational stage?

( ) yes

( ) no

3-   How do you consider the operational internship(s) you have completed?

( ) great

( ) good

( ) regular

( ) bad

4-   How do you consider the didactics of the instructors who taught the internship(s) you did?

( ) great

( ) good

( ) regular

( ) bad

5-   Were the assessments carried out on the placement(s) clear and did they assist your learning?

( ) yes

( ) no

6-   Do you think that the lack of training of the instructor of the operational stage for practical instruction interferes with the technical-professional training of the student trainee?

( ) yes

( ) no

7-   Do you think that failures in instructions that are given inappropriately can create operational safety risks?

( ) yes

( ) no

If you wish, record a comment on the subject covered in the questions above.

APPENDIX 2: QUESTIONNAIRE APPLIED TO AIR TRAFFIC CONTROLLER INSTRUCTORS

( ) CINDACTA 1   ( ) CINDACTA 2(   ) CINDACTA 3( ) CINDACTA 4

( ) ICEA(   ) SRPV   ( ) PAM

1-   How long have you been teaching?

( ) 1 year or less

( ) between 1 and 5 years

( ) between 5 and 10 years

( ) + 10 years

2-   Do you have a course of instruction?

( ) yes

( ) no

if yes, which course(s)?

3-   Have you ever taught on an operational stage?

( ) yes

( ) no

If the answer is yes, did you have specific training to start teaching in the internship?

( ) yes

( ) no

4-   Which educational approaches or pedagogical theories you feel safe to work with:

( ) inatistic maturationalist   ( ) technicist

( ) behaviourist   ( ) constructivist

( ) social-interactionist   ( ) domains and levels of learning

( ) none of them, you have your own way of working that is not based on these theories but has always worked

5-   Mark the concept that best defines evaluation for you:

( ) form of, in a training course, reproving those who do not have conditions to exercise the activity

( ) way to identify prerequisites, control student learning and classify them in a group or class

( ) stage that occurs compulsorily in the courses, it is part of the teaching process.

6-   Do you make lesson plans?

( ) yes

( ) no

( ) I do not know how to make a lesson plan

7-   Do you consider yourself to have a profile for teaching?

( ) yes

( ) no

8-   Do you think that everyone in SISCEAB who gives instruction has the profile for it?

( ) yes

( ) no

9-   Based on your experience and making an estimate, how many people in SISCEAB who give instruction are adequately trained by the system to do so?

( ) from 0 to 10%

( ) from 10 to 20%

( ) from 20 to 30%

( ) from 30 to 40%

( ) from 40 to 50%

( ) + 50%

If you wish, record a comment on the subject covered in the questions above.

DOCUMENT ENTRY SHEET

[1] CLASSIFICATION/TYPE	2. DATE	[3] REGISTRATION N°	4. N° OF PAGES
DP	01 February 2013	DCTA/ITA/DP-060/2012	110

[5] TITLE AND SUBTITLE:

The training of practical operational instructors for the preservation of flight safety

[6] AUTHOR(S):

Tereza Cristina Buonocore Nunes

7. INSTITUTION(S)/INTERNAL BODY(IES)/DIVISION(S):

Technological Institute of Aeronautics - ITA

[8] KEYWORDS SUGGESTED BY THE AUTHOR:

Instructor, Air traffic control, Training; Practical instruction

9. KEYWORDS RESULTING FROM INDEXING:

Air traffic control; Instructors; Personnel training; Training programmes; Aeronautical engineering, Teaching.

[10] PRESENTATION: **(X) National**   **( ) International**

ITA, São José dos Campos. Professional Master Course in Aviation Safety and Continued Airworthiness. Graduate Program in Aeronautical and Mechanical Engineering. Advisor: Prol®. Ligia Maria Soto Urbina; joint advisor: Cap. Jorge Augusto Martins. Defense in 23/11/2012. Published in 2012

[11] SUMMARY:

Improving the training of flight controllers is very important to strengthen air transport in Brazil. With the increasing flow of aircraft, it is of utmost importance to have a well-trained professional to perform the function in order to ensure safety in the operation. The training of this professional goes through distinct stages, one theoretical and the other practical. Although closely linked and interdependent, they occur at different times. The practical phase is subdivided into two more phases, simulated practice and real practice. The real one occurs with the professional, still in training, already in his workstation with the accompaniment of a practical instructor, responsible for the accompaniment and training of the trainee who was approved in the previous phases and is already apt for real time training. The practical instructor is of fundamental importance in the formation of the air traffic controller and assumes a huge responsibility when dealing with professionals still in training during the operation. This instructor aims to bring new knowledge to the trainees, make them work with unusual situations, act naturally, objectively and assertively in the exercise of the function, identify possible doubts or learning problems and correct throughout the teaching process, monitoring actions and constantly evaluating the trainee to ensure the quality of the air traffic control professional's training, minimizing risks to aviation with regard to control. The instructor's function is very relevant and requires, on the other hand, an adequate training with specific focus on practical instruction. In this context, the objective of this work is to propose an adequate training for these professionals in accordance with international guidelines. For this purpose, a qualitative research was conducted, mainly using academic and documental literature review to propose a basic curriculum for air traffic controller practical instructor training. The research was structured in order to present and analyze the national and international panoramas regarding the characteristics of the courses offered and content of the practical-operational air traffic controller (ATC) instructor training. A primary data search was also conducted to identify air traffic problems that could be attributed to problems in instruction. Additionally, it was presented and discussed the PHO that gathers the main improvements that have been developed to improve operator instruction. This program is an advance, but we identified the need for the instructor to have previous knowledge so that the instructor's operational qualification is successful. Finally, a chapter is dedicated to propose a set of basic pedagogical contents to be taken into consideration when building the practical instructor qualification, thus forming part of a curricular base able to strengthen the controllers' instructors training.

[12] DEGREE OF SECRECY:

( ) OSTENSIBLE (X) RESERVED   **( ) CONFIDENTIAL( )**   **SECRET**